D0037522

WORMS EAT MY GARBAGE
is the third of a series designed
to help people regain control
over their own
lives. By devel-
oping skills, sharing
knowledge, and
working cooperatively,
we can accomplish together what none of us
could do alone.

Flower
Press

Gillian O'Roarke 467-0085

WORMS EAT MY GARBAGE

Flower Press

Checklist

WORMS
Eat My Garbage

by
Mary Appelhof

with illustrations by
Mary Frances Fenton

Flower Press
Kalamazoo, Michigan USA

Gratitude is expressed to the Kalamazoo Nature Center, Kalamazoo, Michigan, and Dr. Dan Dindal of the State University of New York, Syracuse, New York, for permission to reprint the following:

artist: Mary Frances Fenton

1-2-3 Worm Box first appeared in "Composting Your Garbage with Worms", published by the Kalamazoo Nature Center, 1979.

"Variety of Garbage Eaten by Worms" is adapted from "Kinds of Garbage Eaten by Worms" in *Winter Composting with Worms*, Final Report from the Kalamazoo Nature Center to the National Center for Appropriate Technology, 1979.

artist: Dr. Dan Dindal

"Food Web of the Compost Pile" appeared in *Ecology of Compost*, published by the State University of New York College of Environmental Science and Forestry.

Library of Congress #82-242012

Worms Eat My Garbage. © 1982 by Mary Appelhof. All rights reserved. Printed in the United States of America. No part of this book, including illustrations, may be reprinted, photocopied, memory-stored, or transmitted by any means without written permission from the publisher. Brief quotations embodied in critical articles or reviews are permissible.

First edition.

Published by Flower Press
10332 Shaver Road
Kalamazoo, Michigan 49002
U.S.A.
ISBN 0-942256-03-4

Contents

Preface to the Seventh Printing

This little guide is now in the hands of over 22,000 individuals and institutions in 50 states and as many foreign countries. I interact with hundreds of people who use worms to eat their garbage, too. The collective knowledge of home vermicomposting expands with every new worm bin.

Based upon this welcome sharing, I recommend one change. For the worm bin described in Chapter Three, I suggest corner reinforcements and a hinged lid, as developed by the urban gardening organization, Seattle Tilth. The lid provides a bench to sit upon, discourages access by rodents, and helps to maintain proper moisture conditions.

To make it easier to get started composting garbage with worms, I now offer a worm bin made of recycled plastic with a unique aeration system. I call it my **Worms Eat My Garbage™ Worm-a-way™**. People like the convenience of having the worm bin, book, and worms delivered to their door. This unit is becoming popular in classrooms, as well.

With the increased interest in handling our garbage in environmentally-sound ways, worm bins may well be the wave of the future. May this book be your guide to an effective, enjoyable way to get rid of your garbage and produce your own potting soil.

Mary Appelhof
January, 1992

Acknowledgments

Just as earthworms take raw organic materials, rework them, and combine them with minerals of the earth to make nutrient-laden worm castings, talented generous people have taken my original manuscript, reworked it, and added shape, figure, and form to create this book. I am indebted to all who made the process such a pleasant task. These include Dr. Dan Dindal for content review, comments, support, and illustrations; Dr. Ed Neuhauser for technical review; Ethel Keeney for editing; Dr. Roy Hartenstein for extensive laboratory studies; Cheryl Lyon for the plant growth response experiment; worm bin owners for feedback on what works and what doesn't; Elin Shallcross for compiling the index; Peg O'Neill, Loring Janes, Gail Bosshard, and Sylvia Brown for proof-reading; Mary Ellen Brown for the bio; and Dr. Scott Geller, Dr. Lewis Batts, Dr. Bethe Hagens, Hazel Henderson, and Mary Brown for their generous comments. To Kathe McCleave goes major credit for careful and thorough, but gentle, editing. Sue Janes wove the tapestry of type. And finally, Mary Frances Fenton, whose fine illustrations add so much, and who contributed not only artistic, editing and consulting talent, but also kept the worms bedded, fed, and counted when I was too busy working on the book. I thank you all.

MA August 1982

Printed on recycled paper.

Foreword

Worms eat my garbage. They've been eating it for the past ten years. Before then I used to object to the smell that developed when I dumped coffee grounds, banana peels, table scraps, and other food waste into my kitchen wastebasket. I got rid of the smell in my kitchen when I learned how to compost food waste outdoors by mixing it with grass clippings, leaves, manure, and soil. However, although composting worked fine in the spring, summer, and fall, it was not convenient during a northern winter. Either the ground froze, or I'd have to wade through the snow to get to the compost pile. Sometimes, even the compost pile froze! Now, I let earthworms help with the composting inside the house. And there is less odor when waste goes into the worms instead of the wastebasket.

The process of using earthworms and microorganisms to convert organic waste into black, earthy-smelling, nutrient-rich humus is known as *vermicomposting*. I began vermicomposting when I ordered a small quantity of redworms through the mail and started to experiment with them. I placed them in a container in my basement, provided a bedding for the worms, and buried most of my kitchen waste in this bin. Using about a pound of worms, I buried 65 pounds of garbage during a 110-day period. When spring came, I used the resulting vermicompost in my garden and discovered that the production of broccoli and tomatoes was much better than I had dreamed possible.

The whole vermicomposting process is very simple and has obvious advantages in addition to increasing garden yields. Worms don't make noise and they require very little care. Because newspapers and packaging, cans, and glass never get mixed in with the smelly stuff, they are cleaner and easier to recycle. I have wonderful fertilizer for my plants and I save the money that used to pay for weekly trash pickup. Oh yes, and I always have plenty of worms for fishing.

I learn more about vermicomposting every time I set up a new unit. That practical experience plus a biology-teaching and research background helps me evaluate the published information available. It also convinces me that there is a need for a detailed, yet simple, manual of vermicomposting.

Worms Eat My Garbage is the third and most comprehensive publication I have written describing how to set up a home vermicomposting system. The first was a two-page flyer published by Flowerfield Enterprises in 1973 entitled "Basement Worm Bins Produce Potting Soil and Reduce Garbage." A second brochure, "Composting Your Garbage with Worms," was published by the Kalamazoo (Michigan) Nature Center in 1979 as part of a grant from the National Center for Appropriate Technology. Although several revisions were made in 1981, it is obvious that people want more information than a four-page pamphlet can supply.

Thousands of people, both children and adults, have seen worms at work in demonstration vermicomposting units at energy fairs, harvest festivals, barter fairs, and garden clubs. After the initial "yuch," questions pour out. This book is an attempt to answer the most common questions. I hope it convinces you that you, too, can vermicompost, and that this simple process with the funny name is a lot easier to do than you thought. After all, if worms eat *my* garbage, they will eat yours, too.

1.
What should I call it?

There are those who use the term "home vermicomposting system" instead of "worm bin" because it sounds more sophisticated. They are right on both counts. It is sophisticated and it *is* a system. The system consists of:

1. **A Physical Structure:** the box or container
2. **Biological Organisms:** the worms and their associates
3. **Controlled Environment:** temperature, moisture, acidity
4. **Maintenance Procedures:** preparing beddings, burying garbage, separating worms from their castings, making use of the castings (worm manure)

I hesitate to use "home vermicomposting system" exclusively because the term itself might frighten away some who would feel more comfortable with "worm bin." It sounds a lot less intimidating to just build a wooden box with holes in the bottom, add moistened bedding and worms, bury garbage, harvest worms, and set up fresh bedding as necessary. If calling it a worm bin encourages you to try the technique, then call it a worm bin, by all means.

On the other hand, the system is so complex that there is a lot more that can be learned about it. Just what *is* going on in that box? Are the worms really eating the garbage? Or are they eating the bacteria, protozoa, molds, and fungi that are breaking down the food waste? Can conditions become too acid? How can you tell? What kinds of food might cause overacidity? When is the best time to harvest if you want more worms? When do you harvest to get the best castings? What is the best size container to use for a given quantity of garbage?

If you like to compare notes about the ideal temperature for cocoon production, or acceptable degrees of anaerobiosis, you might want to say you have a home vermicomposting system. I know a lot about vermicomposting. I have such a system. But, I also have a lot more to learn. For myself, I just say, "Worms eat my garbage. Wanna see my worm bin?"

2.
Where should I put a worm bin?

In deciding where to put your worm bin, consider both the worms' needs and your own. That may sound elementary, but I've learned from experience that there are a few basics you should think about in advance. Adjusting now for these items will help determine later how successful and enjoyable a worm bin will be for you.

To make the worms happy, you'll need to think about temperature, moisture and ventilation. Equally as important, to make you happy, you'll want to consider your expectations, convenience and aesthetic preferences.

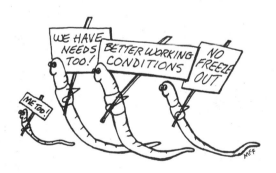

NEEDS OF THE WORMS

Temperature
You will be using redworms (for reasons that I will discuss later). Redworms tolerate a wide range of temperatures, but they should not freeze. They have worked their way through garbage in a basement bin with temperatures as low

as 50°F.* The most rapid feeding and conversion of waste will probably occur at temperatures between 55-77°F. Bedding temperatures above 84°F could be harmful to the worms. The temperature in moist bedding is generally lower than the surrounding air because evaporation of moisture from the bedding has a cooling effect on a well-ventilated location. Locations that could get too hot include a poorly-ventilated, over-heated attic; outside under a hot sun, or higher elevations in a greenhouse.

Earthworms have successfully weathered cold northern winters in pits dug into the ground and covered with manure, straw and leaves which provide heat, food, and insulation. The problem with an outdoor pit for winter garbage disposal is that when the temperature drops to zero, you shouldn't disturb the protective snow covering in order to bury food waste. Such protection will save the worms, but your garbage also piles up!

Moisture

All worms need moisture. They "breathe" through their skin, which must be moist for exchange of air to take place. Since you can always add water to the bedding when necessary, your major concern with moisture is to place your worm bin where there is no danger of flooding, which could cause the worms to drown.

Ventilation

Worms use oxygen in their bodily processes, producing carbon dioxide, just as we do. It is important that you allow air to circulate around the unit. Wrapping it up in a plastic bag, for example, might be tidy, but it would quickly smother the worms.

YOUR NEEDS

To meet your needs, a home vermicomposting system will have to provide a convenient method for converting organic waste to a usable end product. Potential end products are a supply of worms for fishing, worm castings for plants, or vermicompost for use in your garden. *Vermicompost* is a more general term than *worm castings*. A casting is the material deposited after it's moved through the digestive tract of a worm. Vermicompost contains worm castings, but also consists of partially decomposed bedding and organic waste. Worms of all ages, cocoons, and associated organisms may be found in vermicompost. If a worm bin is left untended for six months or so, worms will eat all of the bedding and organic waste, depositing castings as they do so. In time, then, the entire contents will be fully converted to worm castings. Since no food remains for the worms, most worms will die and be decomposed by the other organisms in the worm bin. The few worms that live will be small, inactive and undernourished.

The effectiveness of your vermicomposting system will depend partly upon your expectations and partly upon your behavior. You can reasonably expect to bury a large portion of your biodegradable kitchen waste in a properly prepared worm bin, check it occasionally, make judgments about what must be done, then harvest worms and vermicompost or worm castings after a period of several months. You cannot expect to merely dump all the trash from your kitchen into a worm bin, add some worms, and come back in only two weeks to collect quantities of fine, dark worm castings to sprinkle on your house plants. If this is your expectation, revise it to something more realistic, along the following lines, or don't even begin.

Expectations

The difference between what is and what is not reasonable to expect has to do with the kind of material you bury, the environment you provide for the worms, the length of time you are willing to wait to observe changes, and the character of the end products. It isn't that difficult when you know what you want. These are some guidelines that will help you make reasonable judgments about what you must do to maintain your worm bin so that it is effective based upon your expectations.

Goal: Extra worms, for example, for fishing
Maintenance level: High

Some of you will want to produce more worms than you started with so that you can have a ready supply for fishing. Expect to harvest your bin every two to three months, transfer worms to fresh bedding, and accept vermicompost that is less finished than if you were to leave the worms in their original bedding longer.

Goal: Finished worm castings for plants
Maintenance level: Low

Those who prefer to obtain castings in the most finished form also have the advantage of extremely low maintenance. You will bury food waste in your worm bin for about four months, then leave it alone. You won't have to feed or water the worms for the next few months, but just let the entire culture proceed at its own pace. The worms will produce castings continuously as they eat the bedding and food waste. The disadvantage of this program is that, as the proportion of castings increases, the environment for the worms becomes less healthy. They get smaller, stop reproducing, and many die. As you wait for the worms to convert all the bedding and food waste to castings, you will have to deposit fresh batches of food waste else-where—perhaps in a compost pile. However, your worm bin will provide a quantity of fine castings giving you a homo-geneous, nutrient-rich potting soil. If you prefer this system, you may have to purchase worms every fall when you set up your worm bin. This low maintenance program enables you to vermicompost inside during the cold winter months, compost outside in the traditional manner when the weather warms up, and have finished worm castings from your indoor worm bin sometime during the summer. A name for this maintenance technique might be *"lazy person's."*

Goal: Continuous worm supply plus vermicompost

Maintenance level: Medium

"Middle of the roaders" can opt for a program that requires just enough maintenance to keep the worms healthy. You will harvest fewer worm castings, but you should still have ample quantities of vermicompost to use on your house plants and garden, and enough worms to set up your bin again. About every four months, you will need to prepare fresh bedding and select one of several techniques for separating worms from vermicompost. These techniques are described in Chap. 10.

It should be apparent that the effectiveness of your home vermicomposting system will depend as much upon you as it will depend upon the worms.

Convenience

The convenience of your home vermicomposting system is directly related to its location. There are various possibilities.

Kitchen. Since food preparation is done in the kitchen, the most convenient location for a worm bin might be

there, too. One of my friends has a worm bin on top of his dishwasher with a cutting board serving for a lid. When he is through chopping cabbage, celery or whatever, he just slides the top back and scrapes the waste into his worm bin. You can't beat that for convenience!

Patio. A patio off the kitchen could be an excellent location for a home vermicomposting unit in climates where

freezing temperatures are not a problem. It would be close to the origin of food waste, close to a water supply for maintaining proper moisture, and it would have plenty of ventilation. Just as you can expect to get dirt on the floor when you mix potting soils to repot plants, the periodic maintenance in separating worms from vermicompost can get messy, so doing that on the patio will also keep the dirt outside.

Garage. A well-ventilated garage would be a satisfactory location for a worm bin if it doesn't freeze. It also will provide shade during hot weather.

Basement. Locating a worm bin in a basement, if you have one, has the advantage of keeping it out of the way. If problems develop—there might be short-term odors or occasional fruit flies—the worm bin is not in the immediate living quarters. You might find it inconvenient to have to go downstairs whenever you want to bury garbage, however. Basements do meet the worms' temperature needs, being cooler in summer and rarely freezing in winter. Since they are not in the way for most people, many worm bins, including mine, are located in a basement.

Aesthetic considerations

For some of you, locating your worm bin will depend considerably on what it looks like. If it is a custom-made unit of laminated maple with sturdy legs on ball rollers and looks like a piece out of the Nieman-Marcus catalog, you will want it where you can show it off most readily.

More realistically, before you decide where to locate your worm bin, 1) determine how large your unit must be to process your kitchen waste; 2) assess the space you have available; and, 3) determine whether you want it to be merely functional and out of the way, or whether you want it to be the center of attention. To put it another way, how many guests do you want to have tramping through your basement? Or, how many guests can deal with sitting on a window seat worm bin in the dining room? From my experiences, I can guarantee you, until worm bins are common, almost everyone who visits is going to want to see yours!

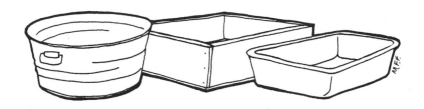

3.
What kind and size container should I use?

A variety of containers make satisfactory worm bins, including wooden boxes, and galvanized metal washtubs or plastic utility tubs. Although specific instructions for building wooden worm bins are given later, general requirements include:

Shape

Your container should be shallow (8 to 12" deep) for three reasons:

1. Redworms tend to be surface feeders.

2. The bedding can pack down in a deep container, compressing the air out of the bottom layers and make it more likely to develop foul-smelling *anaerobic* conditions.

   The greatest concern that people express when they hear about placing kitchen waste in a container to be kept inside the home is, "But won't it smell?" The answer is, "Not too badly *if* it is properly set up and maintained."

   We are trying to create an *aerobic* environment, one in which oxygen is present throughout the bedding. Oxygen is necessary not only for the worms, but for the millions of microorganisms that are also breaking down the food waste. Among the end-products of their bodily processes are carbon dioxide and water, neither of which smells.

Some microorganisms live and reproduce only when no oxygen is present; these are *anaerobic*, or decomposers requiring no oxygen. As they break down waste, they produce gases that have foul-smelling, disagreeable odors.

The secret to having an odor-free worm bin is to have oxygen available throughout the bedding so that both the worms and microorganisms can break down the waste aerobically.

3. Given bins of different shape but equal volume, the one with greater surface area provides better aeration and more locations to bury waste on a rotating basis.

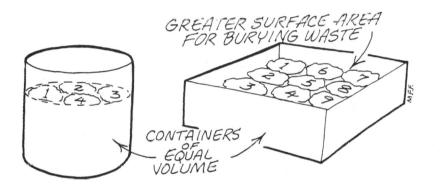

Size

How do you decide how big your home vermicomposting unit should be? First, you need some idea of how much organic kitchen waste you produce. You may want to keep track of how many pounds you throw away each week. Is it 5, 10, or 15 pounds? Many things affect this, including how many individuals you have in your household, whether you are a vegetarian (more worm food!), how often you eat out, whether you use prepared mixes or start from scratch, and how often you have to throw away left-overs and spoiled food.

Example 1. My household of two non-vegetarian adults produces about 3.5 pounds of worm food per week. It consists of such wastes as potato peels, citrus rinds, outer leaves of lettuce and cabbage, tea bags or herb tea leaves, moldy leftovers, plate scrapings, cucumber rinds, pulverized egg shells and onion skins. We eat lunch at home only on weekends, and eat dinner out once or twice a week. Our bin is 2' x 2' x 8", or about one square foot of surface for each pound of garbage per week.

Example 2. A one-adult household where much food preparation was done at home had from 1.75 to 12 pounds of waste to feed worms per week. The average over a 14 week period was 5.2 pounds per week. A 1' x 2' x 3' bin with a six square foot surface area was used. Again this approximates one square foot surface for each pound of garbage per week.

Two attitudes affect the volume of waste thrown away. 1) We are more conscious of how wasteful we are, so we tend to be more careful and have less to throw away. 2) We know that the waste will go to good use because the worms will convert it to worm castings which are then used to grow better vegetables, so we feel less guilty about discarding food waste. I don't know how this contradiction actually influences the volume of waste.

Material

If you want to improvise with containers on hand, be sure that the one you select has not been used to store chemicals, such as pesticides, which may kill the worms. Some worm growers suggest that new plastic containers should be scrubbed well with a strong detergent, then carefully rinsed prior to growing worms in them.

Wooden boxes. The most common sizes for wooden boxes are 1' x 2' x 3' (1-2-3 box) and 2' x 2' x 8" (2x2 box). The 1-2-3 box is large, one that should be able to handle an average of 6 pounds of garbage per week, or food waste from a family of from four to six people.

1-2-3 BOX (4 to 6 person box)
Materials needed for the 1-2-3 box are:
2 pieces 5/8" CDX plywood (35-5/8" x 12")*
2 pieces 5/8" CDX plywood (23-3/8" x 12")
1 piece 5/8" CDX plywood (24" x 36")
38 2" ardox nails**, hammer, drill with 1/2" bit

*CDX plywood is exterior grade, good one side; #2 pine boards or scrap lumber can be substituted.
**Ardox nails have a spiral shape which increases their holding power, particularly important for wood which is alternately wet and dry.

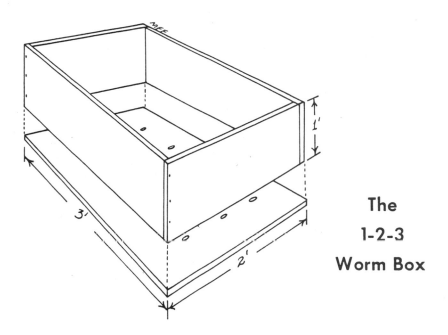

Figure 1. Construction diagram for 1-2-3 Worm Box showing side overlap on corners and holes in bottom for aeration.

Figure 1 shows how to interlock the corners for greater strength. Once the sides are nailed together with about four nails per side, secure the bottom to the sides using five to seven nails per side. Drill twelve 1/2" holes in the bottom for aeration and drainage. These bottom holes require you to place the bin on boards, legs, or casters to allow air to circulate underneath. Interestingly, the worms don't usually crawl out of the holes. However, small amounts of bedding or worm castings fall out, so you may want to have a sheet of plastic or a tray underneath your bin.

> **2x2 BOX** (2 person box)
> Materials needed for the 2' x 2' x 8" box suitable for one or two people are:
> 4 pieces 5/8" CDX plywood (23-3/8"x8")
> 1 piece 5/8" CDX plywood (24" x 24")
> 36 2" ardox nails, hammer, drill with 1/2" bit
> (See 1-2-3 Box notes for ardox nails or substitute woods.)

Nail the sides together, overlapping the corners as shown in Figure 1. Secure the bottom to the sides using about five nails per side. Drill nine 1/2" holes in the bottom for aeration and drainage, and follow the recommendations above for a tray and for raising the box from the surface upon which it sits.

Fancy boxes. Some people have built worm bins as a piece of furniture, providing legs on casters, staining and finishing boards of finer grades of wood, or using a more attractive (and more expensive) grade of plywood, such as birch. Two precautions you should be aware of are to 1) use exterior grade plywood, since the box will be damp most of the time, and you don't want the layers to separate from each other, and 2) avoid highly aromatic woods, such as redwood or cedar, which may be harmful to the worms.

How long will a wooden box last? Used continuously, without ever letting the box dry out, unfinished wooden boxes should last two to three years. Longevity can be increased by letting the box dry out for several days between set-ups. Some people prefer to rotate between two boxes for more convenient maintenance and to allow the boxes to dry. A good finish that seals all edges, such as polyurethane varnish, epoxy, or other waterproofing material, should extend box-life considerably.

4.
What are worm beddings, and where do I get them?

A major component of your home vermicomposting system is bedding. Worm beddings are functional since they not only hold moisture, but provide a medium in which the worms can work, as well as a place to bury the garbage. Worm bedding is usually cellulose in some form. It must not be toxic to the worms, since they will eventually consume the bedding as well as the garbage. The most desirable beddings are light

and fluffy, conditions necessary for air exchange throughout the depth of the container. This helps control odor by reducing the chances that anaerobic conditions will develop.

If left six months or more, all the bedding may be converted to worm castings. It can become so dense that the worms have a hard time moving through it. When these castings are allowed to partially dry and are then screened, it is impossible to identify either bedding or the garbage that was originally buried. However, normal procedures for maintaining a healthy worm population (discussed in Chap. 10), require that worms be removed from the bedding before it is completely converted to castings.

Many materials make satisfactory beddings. Some of the more common beddings are listed below, along with brief comments on their advantages and disadvantages. Your choice can be highly individual, depending upon availability, convenience, and economic considerations.

Shredded corrugated (cardboard)

Shredded corrugated cartons make one of the best beddings for home vermicomposting systems. I purchase this material in wholesale quantities from a company that shreds scrap cuttings from corrugated carton manufacture. Although shredding equipment similar to that used in pulverizing newsprint for insulation is used, the fire-retardant chemical added to insulation would contaminate worm bedding. For this reason, machinery separate from that used to produce insulation is required for worm bedding. Because minimum order requirements, warehousing, marketing, and distribution costs require a considerable investment, the availability to retail purchasers of shredded corrugated suitable for worm bedding will depend greatly upon demand.

Although light, fluffy, shredded corrugated is easy to dampen to the proper moisture content, be careful when handling it. Before the shredded material is moistened, fine, powdery dust is easily raised. As with any quantity of fine, particulate material, it may be harmful to breathe.

ADVANTAGES
- Clean
- Odorless
- Easily prepared
- Makes good
  castings

DISADVANTAGES
- Must be purchased
- May be hard to find
- Dries out on top and edges
- Dust can be harmful to
  breathe

Machine-shredded newsprint or computer strips

Paper shredding machines produce high volumes of good worm bedding from computer print-out paper and from newsprint. The long, tangled lengths of quarter-inch wide strips are easily moistened, and there isn't the dust problem there is with shredded corrugated. The strips, however, don't hold moisture as well, since they provide more surface area from which water can evaporate. The worms will eventually eat the softened paper, as they do the corrugated though, so the end result will also be worm castings.

The most commonly asked question about newsprint is, "Isn't the ink harmful to the worms?" No. The basic ingredients of black ink are carbon black and oils, neither of which is toxic to worms. While black ink is not a problem, colored inks are more of a concern. At one time heavy metals, such as lead and chrome, were a major component of their pigment. Recent government regulations, however, now prohibit the use of heavy metals in any ink used for products that may be eaten by youngsters, such as the funny papers. While the feds are more concerned about our kids than they are about worms, usually it's safe now to say that ink from newsprint will not kill your worms.

Where can you get paper that is shredded? Ask around. Banks and universities frequently shred volumes of computer records; many offices have paper shredders. One hobbicraft business bought a machine to shred newspapers so that its customers would have packing for their greenware and finished ceramic pieces. You only need a supply a few times a year.

ADVANTAGES
• Clean
• Usually free for the hauling
• Odorless
• Easily prepared
• No dust

DISADVANTAGES
• May be hard to find
• Dries out more readily than shredded corrugated

Hand-shredded newsprint

The least expensive and most readily available bedding is newspaper strips you shred by hand. By fully opening a section of newspaper, tearing it lengthwise down the center-fold, gathering the two halves, tearing it lengthwise again, and repeating the process five to six times for each section, you will get strips ranging from one to three inches wide. It doesn't take long to fill your bin with bedding and have enough on hand in a large plastic bag to change the bedding in a few months.

ADVANTAGES
• No cost
• Readily available
• Odorless
• No dust

DISADVANTAGES
• Requires preparation time
• Inked newsprint can be dirty to handle
• Large strips dry out more readily than shredded corrugated or machine-shredded paper
• Tends to mat down in layers, making it difficult to bury garbage

Animal manures

Composted horse, rabbit or cow manure is good bedding for worms, since that is a natural habitat for them. Manure may be difficult to obtain and should not come from recently wormed animals. Some people may object to the initial odor, although that will disappear within a few days after worms are added. Manure is likely to have other organisms such as mites, sowbugs, centipedes, or grubs, which some people would rather not have in their homes.

Setting up a bin with manure is a two-stage process, since the medium may heat up beyond the tolerance of worms when water is added. However, the worms can be added in about two days when it cools down. Manure can be mixed with peat moss to make it lighter and less compact.

ADVANTAGES
• Can be free for the hauling
• Natural worm habitat
• Variety of nutrients present
• Makes good castings

DISADVANTAGES
• Initial odors can be objectionable
• Unwanted organisms may be present
• May initially heat up, delaying time when worms can be added
• Can get compacted

Leaf mold

The bottom of a pile of decaying leaves can yield a satisfactory bedding in the form of decomposed leaves. If they are wet, you'll probably even find some worms! Maple leaves are preferable to oak, which take a long time to break down. Recent concern has been expressed about the high concentration of lead in leaves obtained from heavy traffic areas. Lead in the exhaust from cars using leaded gasoline settles on the leaves, so be cautious in using many leaves from such a source.

ADVANTAGES
• No cost

• Natural worm
  habitat

DISADVANTAGES
• Unwanted organisms may
  be present
• Leaves can mat together,
  making it difficult to bury
  garbage

Peat moss

    Canadian peat moss is a standard bedding among some commercial worm growers. It has superior water retention characteristics, but provides almost no nutrients for the earthworms. It is highly acid, and other substances harmful to the worms may be present if peat moss is not leached out by soaking it in an excess of water for several hours prior to squeezing it out for bedding. It has also become more expensive in recent years.

    Peat moss can be effectively mixed with any of the other beddings described to aid in water retention or to make manure beddings less dense. I recommend using one-third to one-half peat moss.

ADVANTAGES
• Clean
• Odorless
• Retains moisture
  well
• Good for mixing
  with other
  beddings

DISADVANTAGES
• Must be purchased
• Few nutrients for worms
• Acidity can be a problem

    Additions to bedding. You may have noticed that I have not mentioned using dirt or soil for bedding. In nature, redworms are litter dwellers; that is, they are found among masses of decaying vegetation such as fallen leaves, or manure piles, or under rotten logs. They are present in mineral soils only when large amounts of organic materials are present. Although one investigator recommends using a thin (1/2") layer of soil in a container holding worms, I have not found it to be essential for home vermicomposting systems. A big disadvantage is weight. With just a half-inch of soil, for example, a container is extremely heavy. I will, however, usually add a handful or two of soil when I initially prepare the bedding. This provides some grit to aid in breaking down food particles within the worm's gizzard.

Powdered limestone (calcium carbonate) can also be used to provide grit. It has the further advantages of helping to keep conditions in the bin from becoming too acid, and providing calcium, which is necessary for worm reproduction. Since pulverized egg shells serve the same purpose, and I add them regularly, I frequently don't bother with the limestone. **CAUTION:** Do not use slaked, or hydrated lime. Use the kind that can be mixed with feed, or that is used to line athletic fields. The wrong kind of lime will cause your worms to react violently and will kill them.

5.
What kind of worms should I use?

Why redworms?

Redworms are the most satisfactory kind to use in your home vermicomposting system, for a number of reasons. They process large amounts of organic material in their natural habitats of manure, compost piles, or decaying leaves. They reproduce quickly and in confinement. When small organisms are raised in a controlled environment, they are said to be *cultured;* the culture of earthworms is known as *vermiculture.* Because sufficient markets exist encouraging people to culture redworms on a part or full-time basis, it is possible to purchase them almost any season of the year. They can readily be shipped via package delivery services or through the mail.

Other common names. Calling earthworms by common names can cause problems in communication. What I call "redworm," you may know as "red wiggler." Your neighbor may call it "manure worm." The bait dealer down the road may refer to it as "red hybrid." Other common names for this same animal are: fish worm, dung worm, fecal worm, English red worm, striped worm, stink worm, brandling, and apple pomace worm. Redworms frequently have a pronounced pattern of alternate red and buff stripes, which may explain why they are called "tiger worms" in Australia. With so many names, how can any of us know when we are talking about the same worm!

Scientific names

To be certain they are talking about the same thing, scientists have developed a precise system for naming organisms. Since much information in this book comes from scientific papers, I will be using scientific names. So that you won't be confused when I do, here are some basic rules that all scientists follow:

The name of each organism consists of two words, the first of which is called a *genus* (plural: *genera)*, the second, the *species*. All organisms of the same genus are more closely related than those of different genera. Human beings are members of the species *Homo sapiens*. Correct usage requires that the genus name always be capitalized, the species name be lower case. Both terms are of Latin or Greek origin, or they are Latinized. Scientific names are italicized in print, or if typewritten, underlined.

The redworms I use are *Eisenia foetida* (which I pronounce i see' nee a fet' id a). Another redworm that could possibly be used is *Lumbricus rubellus*, which also inhabits compost heaps and manure piles. *L. rubellus* is more likely to be a soil dwelling earthworm than *E. foetida*, but the soil would have to contain large amounts of organic material.

Nightcrawlers

Most people recognize the nightcrawler, known to some as the dew worm, night walker, rain worm, angle worm, orchard worm, or even night lion. Scientists call it *Lumbricus terrestris*, and it is unquestionably the most studied of the 3000 species of earthworms currently named. A recent bibliography of earthworm research lists over 200 scientific studies with data on *Lumbricus terrestris!*

Nightcrawlers are not a suitable worm for the type of home vermicomposting system described here. I once placed 80 nightcrawlers in my worm bin along with the redworms already there. Two months later, I found only one live night-crawler, and that was immature. Although satisfactory environments can be created for nightcrawlers indoors, they require large amounts of soil, and the bed temperature cannot exceed 50°F. Indoors, your box temperature is likely to get higher than that. Nightcrawlers dig burrows and don't like to have their burrows disturbed. If you try to bury garbage, nightcrawlers move quickly around the surface of the box trying to escape your digging.

Nightcrawlers do play important roles in soil fertility. These large soil-dwelling earthworms have extensive burrows extending from the ground surface to several feet deep. They come to the surface on moist spring and fall nights and forage for food, drawing dead leaves, grass, and other organic material into their burrows where they feed upon it at a later time. Nightcrawlers perform important soil mixing functions. They take organic material into deeper layers of the soil, mix it with subsoils that they consume in their burrowing activities, and bring mineral subsoils to the surface when they deposit their casts as dark globs of coiled earth near the burrow entrance. Through their burrows, nightcrawlers also aid in soil aeration and in water retention by increasing the rate at which water can penetrate the deeper soil layers. They may not be good for your worm boxes, but they are very good for your gardens.

Garden worms

I can't tell you what kind of worm you will find if you dig in your garden. Scientists describe several soil and pasture dwelling species including the *Allolobophora caliginosa* complex of species including *Allolobophora chloritica* and *Aporrectodea turgida* and *A. tuberculata*, among others. To identify these species one would need to count the number of segments, identify the type of projection over the mouth, locate the position of the various openings for sexual organs, and identify the pattern for the setae (bristles) on each segment. With suitable magnification, a good pair of forceps, two or three well-illustrated books, and a considerable struggle, I might be able to come up with a tentative identification in a half-day's time. To be certain, I would send a properly pre-served specimen off to a good earthworm taxonomist. Identifying earthworms is not for the faint of heart.

6.
What is the
sex life of a worm?

Before determining how many redworms you need to start vermicomposting, an understanding of their amazing reproductive potential is helpful. Two of the reasons for using redworms are that they reproduce quickly and in culture. When some people learn how fast redworms reproduce, they become concerned that redworms will "take over the world," but their numbers are controlled by environmental factors.

Many people know that an earthworm has both sexes, and may wonder why mating is even necessary when each worm produces both eggs and sperm. Knowledge of a worm's structure helps explain this.

The swollen region about one-third of the distance between the head and tail of a worm is the *clitellum*, sometimes known as the girdle or band. The presence of a clitellum indicates that a worm is sexually mature. Bait worms with this structure are commonly called "banded breeders," so they are old enough to breed and produce offspring.

Just as worm species differ in external characteristics, they differ somewhat in mating behavior. For example, nightcrawlers extend themselves from their burrows to seek another nightcrawler with which to mate. Attracted by glandular secretions, they find each other and lie with their heads in opposite directions, their bodies closely joined. Their clitella secrete large quantities of mucus that forms a tube around each worm. Sperm from each worm move down a groove into receiving pouches of the other worm. The sperm, in a seminal fluid, enter the opening of sperm storage sacs where they are held for some time.

Redworms differ from nightcrawlers by mating at different levels in their bedding, rather than just upon the surface.

Under proper conditions, they can also be observed mating at any time of year, whereas some species mate only during particular seasons.

Some time after the worms separate, the clitellum secretes a second substance, a material containing albumin. The albuminous material hardens on the outside to form a cocoon in which eggs are fertilized and from which baby worms hatch. As the adult worm backs out of this hardening band, it deposits eggs from its own body and the stored sperm from its mate. Sperm fertilize the eggs inside this structure, which closes off at each end as it passes over the first segment. Sometimes called an egg case, this home for developing worms is more properly called a cocoon.

Cocoons are lemon-shaped objects about the size of a matchhead or a small grain of rice. They change color as the baby worms develop, starting as a luminescent pearly white, becoming quite yellow, then light brown. When the hatchlings are nearly ready to emerge, cocoons are reddish. By observing carefully with a good hand lens it is sometimes possible to see not only a baby worm, but the pumping of its bright red blood vessel. The blood of a worm is amazingly similar to ours, having the same function of carrying oxygen, and having iron-rich hemoglobin as its base.

It takes at least three weeks development in the cocoon before one to several baby worms hatch. The time to hatching is highly dependent upon temperature and other conditions. (I worked with worms for years before I ever saw a worm emerge from its cocoon. I have observed hatchlings work their way out of their cocoon, thrashing about vigorously. When I turned on bright lights to try to photograph them, they quickly retreated, reacting negatively to light just as adult worms do.)

Newly emerged worms are whitish and nearly transparent, although the blood vessel causes a pink tinge. They may be from one-half to nearly an inch long when they hatch, but they weigh only two to three milligrams. At that size, it would take over 150,000 hatchlings to make one pound of worms.

Although each cocoon may contain as many as 20 fertilized eggs, normally only two to three hatchlings emerge. Because the number of hatchlings varies depending upon such

EARTHWORM MATING AND COCOON FORMATION

EACH WORM HAS <u>BOTH</u> OVARIES AND TESTES.

AREA OF OVARIES AND TESTES

CLITELLUM (BAND)

TWO WORMS JOIN BY MUCUS FROM THEIR CLITELLA. SPERM THEN PASS FROM EACH WORM TO THE SPERM STORAGE SACS IN THE OTHER WORM.

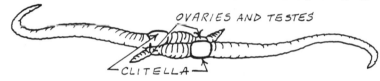

OVARIES AND TESTES

CLITELLA

LATER, A COCOON FORMS ON THE CLITELLUM OF EACH WORM. THE WORM BACKS OUT OF THE HARDENING COCOON.

COCOON FORMATION

EGGS AND SPERM ARE DEPOSITED IN THE COCOON AS IT PASSES OVER OPENINGS FROM OVARIES AND SPERM STORAGE SACS.

APPROX. 1/8" LENGTH

AFTER BEING RELEASED FROM THE WORM, THE COCOON CLOSES AT BOTH ENDS. EGG FERTILIZATION TAKES PLACE IN THE COCOON.

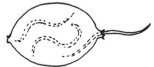

TWO OR MORE BABY WORMS HATCH FROM ONE END OF THE COCOON.

MEE.

factors as the age of the breeder which deposited the cocoon, its nutritional state, the temperature, and whether the temperature is constant or fluctuates daily, it is possible to establish conditions for greater hatchling production.

The time it takes for a baby worm to become a breeder varies, depending on the same factors—temperature, moisture, food availability, and population density. A redworm can be sexually mature and produce cocoons in four weeks, but six is more common. Once it breeds and begins laying cocoons, it can deposit two to three cocoons per week for from six months to a year. Conservatively, then, if a two-month-old breeder laid two cocoons a week for 24 weeks, and two hatchlings emerged from each cocoon, one breeder would produce 96 baby worms in six months (2 cocoons x 24 weeks x 2 hatchlings).

The situation is more complicated than that, however. Before the first two months are up, the first hatchlings will be able to breed. These could produce two cocoons for 16 weeks with two hatchlings coming from each of the four worms resulting from the original breeder's first week's production, or 256 more worms before the six month period is up. The math quickly gets complicated, but since optimal conditions for such geometric increases in numbers will never be achieved, theoretical projections are more confusing than informative. Dr. Roy Hartenstein of Syracuse, New York, has calculated that eight individuals could produce about 1500 offspring within six months time.*

Population controls

With the reproductive potential described, we come back to the question of why worms don't take over the world. Three basic conditions control the size of a worm population. They are: 1) availability of food; 2) space requirements; and 3) fouling of their environment.

When food waste is fed regularly to worms in a limited space, the worms and associated organisms, both microscopic and larger, break down this waste. They use what they can and

*This is based upon two cocoons per worm per week, of which 82% hatch, averaging 1.5 hatchlings per cocoon, reproducing between 5 to 6 weeks, and continuing for 40 to 50 weeks of fertility at 77°F.

excrete the rest. As the worms reproduce, the voracious young worms compete with their parents and all the other worms in the culture for the limited food available. Additionally, all the worms excrete castings, which have been shown to be toxic to members of their own species.

As time goes on, more worms compete for the limited food, and more and more of the bedding becomes converted to castings. The density of the worms may exceed that favorable for cocoon production, and reproduction slows down. Undesirable conditions in their continually changing environment may cause some worms to die. Interestingly, you will rarely see dead worms because they are rapidly decomposed by other associated creatures in this active composting environment.

The controls you exert over your worm population will affect this whole process. You may choose to feed an ever increasing population more and more food. If you want more and more worms, you will eventually have to provide them more space and fresh bedding, and enable them to get away from high concentrations of their castings. You may even choose to become a worm grower, and try to keep up with the ever increasing demand for food, space, and elimination of accumulated end products. But that's another story, and it's more complicated than simply keeping enough worms alive to process your kitchen waste so that you can use the rich end product to grow healthier vegetables and house plants.

7.
How many worms do I need?

With some idea of how fast redworms reproduce, you might conclude that you would eventually produce enough worms to handle all your food waste regardless of how many worms you used initially. You could start with a few dozen redworms and regularly feed them small amounts of garbage. If patience is one of your virtues, you could wait until their natural tendency to reproduce under proper conditions yields several thousand worms. However, most of you will want your worm composting system to handle most of your organic kitchen waste from the day you set it up. You'll want to start with enough worms to consume all the garbage you will be feeding them.

Base the number of worms initially required on two factors: 1) average amount of food waste to be buried per day, and 2) size of bin. Since your bin size will be based upon how much garbage you expect, the amount you will be burying is really the critical factor.

Worm to daily garbage ratio

The relationship between weight of worms required and a given amount of garbage can be expressed as "worm: garbage ratio" (: means "to"). I usually recommend a worm:garbage ratio of 2:1, based upon the initial weight of worms and the average daily amount of garbage to be buried. Thus, if you generate seven pounds of garbage per week, that would be an average of one pound of garbage per day.

$$\frac{7 \text{ lb garbage per week}}{7 \text{ days in one week}} = 1 \text{ lb garbage per day average}$$

For this quantity of garbage, use two pounds of worms initially in a container with about seven square feet of surface

area (the 1-2-3 Box described in Chap. 3 would be sufficient).

Calculations for a household that produces a smaller amount of garbage are:

$$\frac{3.5 \text{ lb garbage per week}}{7 \text{ days in one week}} = 1/2 \text{ lb garbage per day average}$$

Following the suggested worm: garbage ratio of 2:1, use one pound of worms to the one-half pound of daily garbage. Referring back to Chap. 3, your container size should provide about four square feet surface area, or one square foot for each pound of garbage per week.

Quantities of worms are specified in terms of "pounds" rather than "numbers" for a couple of reasons, one personal and the other biological. The personal reason is that the first season I sold worms, my partner and I sorted and counted 50,000 worms—one by one by one. If you ever have to count 50,000 of *anything* one by one, you'll find an easier way to do it, too. From that time on, I have sold worms by the pound. The first question I always get, of course, is "How many worms are in a pound?" Although the number will vary depending upon the size of the worms, there are some guidelines.

Worm growers commonly estimate that there are about 1000 breeders per pound for young redworms. A bait dealer selling worms for fishing would prefer that the worms be considerably larger than that. If redworms don't run between

600 to 700 per pound, customers complain that they are too small to get on a hook.

Many growers sell what they refer to as "pit-run" redworms, or run-of-pit. These are worms of all sizes and ages, from which bait-sized worms may or may not have been removed. Since there could be between 150,000 to 200,000 hatchlings in a pound, the number of pit-run worms in a pound will vary tremendously. However, 2000 is a figure commonly used.

The biological reason for using weight of worms in determining how many to use to start a home vermicomposting system is that it is the *biomass* of the worms that is important, not the number. Worms have been shown to consume more than their weight each day, regardless of their size. Since many small worms can move as much material through their intestines as fewer large worms of equal mass, think in terms of an earthworm biomass sufficient to do the job.

Breeders or pit-run?

There are no hard and fast rules to tell you whether to start with breeders or pit-run. Breeders will lay cocoons more quickly and increase the number of individuals sooner, but they usually cost more from commercial growers because of the labor required to sort them. Also, some growers think breeders take longer to adjust to new culture conditions than do pit-run worms.

If you can order pit-run by the pound, you will certainly get more worms than if you purchase breeders by the pound. These young, small worms will grow rapidly and soon be able to reproduce. If they adjust to their new home faster than breeders would have, you will be ahead starting with pit-run, especially if they are cheaper.

Whichever you start with, breeders or pit-run, when they produce more worms than the garbage you are feeding them will support, many will get smaller, some will slow reproduction, and others will die. Eventually, no matter how many worms you start with, the population will stabilize at about the biomass that can be supported by the amount of food they receive.

Sources of redworms

One way to obtain the pound or two of redworms

that you need to set up your home vermicomposting system is to order them from one of the commercial earthworm growers who advertise in the classified ads of gardening and fishing magazines. Redworms are easily packaged and shipped through the mail, or package delivery systems. Some growers advertise and ship all year-round, others seasonally. As with growing conditions, temperature extremes should be avoided, but if the temperature is colder than 10° or above 90°F, growers will usually wait until the temperatures moderate before they ship the worms.

You may be able to find local growers to provide your initial stock. If you buy redworms from bait dealers, however, expect to pay about twice as much as you would if you buy directly from a grower.

Those of you who like adventure may be able to collect redworms from their natural habitat. Your chances increase if you have a friend with horses, a barn, and a manure pile. You may have to turn over a lot of manure to find any, but then again, you could get lucky and find hundreds in a few pitch-forks full of well-aged, moist manure.

8.
How do I
set up my worm bin?

When you have completed tasks one through five on the checklist that appears on the flyleaf of this book, you are ready to set up your worm bin. You have determined approximately how many pounds of kitchen waste you will dispose of per week, sized and built your container accordingly, selected and obtained your bedding, and ordered or collected your worms. If all materials are on hand, it takes about an hour to set up your bin.

Preparation of worm bedding

Needed:
- Completed worm bin
- Bedding
- 1 to 2 handfuls of soil
- Bathroom or utility scale
- Gallon jug
- Large clean plastic or metal garbage can for mixing bedding

The major task remaining to set up your worm bin is to prepare the bedding for the worms by adding the proper amount of moisture. A worm's body consists of approximately 75 to 90% water, and its surface must be moist in order for the worm to "breathe." By preparing bedding with approximately the same moisture content (75%) as the worm's body, the worm doesn't have to combat an environment that is either too dry, or too moist.

When using shredded corrugated, paper, and/or peat moss beddings, a 75% moisture content can be easily obtained since the residual moisture present is minimal. Just weigh the bedding, and add three times as many pounds of water as you have pounds of dry bedding. To get 75% moisture, for example, add twelve pounds of water to four pounds of shredded corrugated bedding. Or, expressed another way:

water:bedding ratio = 3:1 by weight

As a guideline to determine the amount of bedding you need, it takes from four to six pounds of dry bedding to set up a 2x2 Box, and from nine to fourteen pounds for the 1-2-3 Box. If you don't have a household utility scale, stand on a bathroom scale, first alone and then with your plastic bag full of dry bedding. The difference between the two weights, of course, is the weight of the bedding. To determine the amount of water to add, remember that "a pint's a pound the world around," so a gallon of water, which is eight pints, will weigh eight pounds.

Place about one-half of the bedding into the large mixing container, gently, so the dust won't fly. Add about one-half of the required amount of water, and mix it into the bedding. Then add one to two handfuls of soil, and the remaining bedding and water. Mix again until the water is well distributed throughout the bedding. Now dump the entire contents of the container into your worm bin and distribute it evenly. (The bedding absorbs the water so that little, if any, leaks from the holes in the bottom of the bin.) Your bin is now ready for the worms!

Manure bedding. If you are using manure for bedding, it is more difficult to determine how much moisture to add to obtain the proper moisture content since you don't know how much moisture is already in the manure. Basically, you want the manure damp, but not soggy. If you squeeze a handful and produce three to four drops of water, it's probably all right; twenty drops or a stream of water is too wet.

With manure beddings, remember to add water at least two days before you add worms. Then, if the manure heats up as it begins to compost, the worms won't die from the heat.

Adding the worms and garbage

When your bedding is ready to receive the worms, open their container and dump the entire contents on top of your freshly prepared bedding. Gently spread any clumps of worms around the surface. Leave the room lights on for awhile. The worms will gradually move down into the bedding as they try to avoid the light. Within a few minutes, the majority of worms will have disappeared into the bedding. If any remain on the surface after an hour, assume that they are either dead or unhealthy. Remove them.

Once the worms are down, you may start burying garbage. Place a sheet of plastic, slightly smaller than the bed surface, on top to retain moisture. I use black plastic to keep out light. The worms work up to the surface, and when I lift the plastic, I can see them scramble down into the bedding.

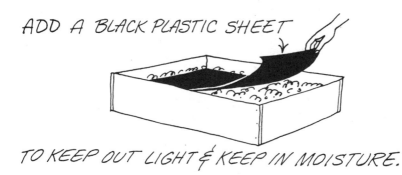

ADD A BLACK PLASTIC SHEET

TO KEEP OUT LIGHT & KEEP IN MOISTURE.

A note about worms prepared for shipping

Most growers package worms for shipping in peat moss, although other materials are used. Experienced shippers pack worms in a fairly dry bedding for two good business reasons. Shipping costs are great, and there is no point in paying to ship excess amounts of water. It is more important, however, to provide a satisfactory environment for the worms. Although worms need bedding with some moisture in it, too much moisture can intensify the effects of temperature extremes during shipping. In mid-summer when the temperature (°F) is likely to be in the 80's or 90's, a drier bedding acts as insulation, plus provides sufficient oxygen for the worms. Too much moisture fills air spaces and the additional heat stimulates natural microorganisms associated with the worms to use up all available oxygen before the worms can get it. If they die, neither you nor I would want to open the box for the smell!

The insulation effect of a drier bedding for packaging also pertains to cold weather shipments of worms. Although the worms will lose some of their moisture to the bedding, they are better off than if they were to freeze because it was too moist and too cold.

If you receive worms that seem dry, assume that the worms will quickly regain their lost body moisture when they are placed in a properly prepared bedding. This should be done within a day or two. Responsible growers try to do what's best for the worms, guarantee their shipments, and provide information so that the customer knows what to expect.

9.
What kind of garbage, and what do I do with it?

What's garbage to me may be trash to you, and slop for the pigs, or food for the dog to someone else. I have previously used such terms as organic kitchen waste and table scraps. It's time to be more specific about what waste you can expect to feed to your worms.

Vegetable waste

Any vegetable waste that you generate during food preparation can be used, such as potato peels, grapefruit and orange rinds, outer leaves of lettuce and cabbage, celery ends, and so forth. Plate scrapings might include macaroni, spaghetti, gravy, vegetables, potatoes. Spoiled food from the refrigerator, such as baked beans, moldy cottage cheese, and leftover casserole also can go into the worm bin. Coffee grounds are very good in a worm bin, enhancing the texture of the final vermicompost. Tea leaves, and even tea bags and coffee filters are suitable. Egg shells can go in as they are, and I have found as many as 50 worms curled up in one egg shell. Usually, I dry them separately, then pulverize them with a rolling pin so they don't look quite so obvious when I finally put vermicompost in my garden.

The list in Figure 2, showing some of the variety of food waste that can be fed to worms, was developed from waste actually buried in worm bins at a nature center. Coffee grounds don't appear on the list merely because none of the six participants' families drank coffee. Use this list as a guideline only. It is not, by any means, comprehensive.

Meat waste and bones

You will not find any meat on the list. When designing the project we excluded the burial of meat for these reasons:

VARIETY OF FOOD WASTE FED TO WORMS

Apples	Grits
Apple Peels	Lemon
Baked Beans	Lettuce
Banana Peels	Malto-Meal
Biscuits	Molasses
Cabbage	Oatmeal
Cake	Onion Peel
Celery	Orange Peel
Cereal	Pancakes
Cheese	Pears
Corn Bread	Pineapple
Cream Cheese	Pineapple Rind
Cream of Wheat	Pizza Crust
Cucumber	Potatoes
Deviled Eggs	Potato Salad
Egg Shells	Ralston
Farina	Tea Leaves
Grapefruit Peels	Tomatoes
	Turnip Leaves

Figure 2. Actual food waste buried in worm bins during a demonstration project at the Kalamazoo (Michigan) Nature Center. Worms can consume many wastes that do not appear on this list.

1) Decaying meat can produce offensive odors from the break-down of proteins in a process known as *putrefaction*. 2) Mice and rats are more likely to be attracted to a worm bin containing meat. 3) Although bones will eventually be "picked clean" by the worms, their sharp edges can injure your hands when garbage is buried. Bones also look unattractive when vermicompost containing them is used in gardens. Since the demonstration bins were to be located in a public exhibit area and seen by thousands of visitors, it was important to avoid such potential problems.

Some meat can be disposed of in a worm bin. I have buried chicken bones, for example. If I dug too soon into the pocket of bedding containing the bones and decaying meat, the odor was bad, though if I didn't disturb it, I didn't notice it. When I harvested the castings, after neglecting the box for several months, what remained was crumbly vermicompost containing blackened, well-picked bones and smelling like damp rich earth.

One worm grower buried the bones from a community chicken barbeque in his worm bins. He said that it took only three weeks for the bones to be picked clean. Dr. Dan Dindal of the State University of New York (Syracuse) suggests adding a good carbon source (such as sawdust) to meat and bones to reduce decomposition time. He finds that if meat is chopped, ground, and thoroughly mixed with the carbon source, rodents won't even be a problem. He says, "I do this successfully all the time in outdoor piles."

The previous examples indicate that some meat and bones can be successfully composted. Advantages exist for putting some meat scraps and bones into your worm bin. The nitrogen from meat is undoubtedly a plus for the system. Worms require nitrogen in a form they can use. Nitrogen is also required by the microorganisms that do much of the composting, and which are eaten by the worms. Since meat contains protein, which is made up of nitrogen, the elimination of all meat from the system could result in a nutrient deficiency for the teeming organisms that constitute a home vermicomposting system. A further advantage of adding some meat is that there will be more plant nutrients in vermicompost produced by worms consuming a greater variety of materials.

A disadvantage of not putting spoiled meat, meat scraps and bones in your worm bin is that you must then find another way to dispose of them, which I do by burying them in my garden. My personal feeling about burying bones and meat waste in a worm bin is that small amounts are all right. If I had a way to grind bones, I would certainly do so to get rid of them and to increase the value of the nutrients of my vermicompost. If I had a large system, away from my immediate living quarters, all wastes, including meat and bones, would be vermicomposted. Since I am limited to a small system inside, I use learned judgment on the quantities I bury. You'll want to experiment cautiously and learn for yourself what your system can take within the design and demands you place upon it.

No-nos

Since what is obvious to some of us isn't always obvious to everyone, I can suggest some things that don't belong in a worm bin. Avoid plastic bags, bottle caps, rubber bands, sponges, aluminum foil, and glass. Such non-biodegradable materials will stay there seemingly forever. They will clutter up your developing vermicompost and it will look more like trash. I have seen the same red rubber band over a three-year period in a large outdoor pit!

Less obvious, but definitely to be guarded against is letting a cat use your worm bin as a litter box. First of all, cat urine would soon make odor intolerable. Secondly, the ammonia in the urine could kill your worms. But the greatest concern has to do with a disease organism that can be carried in the feces of cats, called *Toxoplasma gondii*. Tiny cysts of this protozoan can be inhaled and harbored in tissues of human beings. Frequently no symptoms are felt, but it is possible that a pregnant woman can transmit this disease to her fetus, bearing a child who has brain damage as a result of the *Toxoplasma gondii* passed on from the cat. Although most cats do not harbor the organisms, any cat owner should be very careful in disposing of cat litter. If you have cats, provide a screen or other device to keep them from using your worm bin as a litter box.

Burying the food waste

How do you feed your worms? Or, how do you place your food waste in the bin? I keep a plastic container next to the sink and collect all of the organic waste that will eventually be fed to the worms. If you don't use a lid, thereby permitting air to get to its contents, you can avoid the odors that will soon develop in a tightly closed container. (See aerobic/anaerobic discussion, page 10.) Some users keep the container tightly covered, removing the lid only when adding fresh waste or burying its contents. These tightly closed containers can get pretty "ripe."

About twice a week, I empty the contents into my worm bin. If I have a lot of waste to get rid of, I empty it more often; if I don't have much, less often. In other words, I don't concern myself with seeing to it that the worms are fed daily, twice a week, or even, weekly. My needs, not the worms', dictate how often the worms get fed.

Because I keep records of how much garbage I bury in my worm bins, I weigh the garbage. As I record the weight, I also check my record sheet to see where I buried garbage the last time. I rotate around the box, placing garbage in different areas in sequence like this:

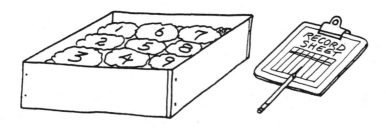

The 2x2 Box I use has about nine locations where I can bury the garbage before I have to repeat. Since I bury garbage about twice a week, four and a half weeks pass before I have to dig into a region with garbage. By then, much of it is no longer recognizable, having been consumed by the worms, or having been broken down by the other natural decomposition processes caused by worm associates in the box.

I cover the newly deposited waste with an inch or so of bedding, and replace the sheet of plastic I have lying loosely on top to retain moisture. With that, I'm through! The whole process takes maybe two minutes, *if* I take the time to poke around looking for cocoons or baby worms.

The worms will tend to follow the waste, but not necessarily when it is fresh. Garbage will undergo many changes as different kinds of microorganisms invade tissues, breaking them down, and creating an environment for other kinds of organisms to feed and reproduce. The worms undoubtedly consume some of the cells from which these tissues are made, but the worms feed also on the bacteria, protozoa, and fungi that thrive in this moist, warm, food-rich environment. Although this book is titled *Worms Eat My Garbage*, I must acknowledge that worms *and* microorganisms eat my garbage. The worms are there because they help keep conditions aerobic and therefore odor-free, reduce the mass of material to be processed, and produce castings far richer than mere compost. Worms don't do the job alone.

Other techniques for adding garbage

Techniques vary. One vegetarian who uses worms to process her kitchen waste has large quantities of peelings, wheat grass roots, and pulp from juicing carrots, celery, and other vegetables. Her container is a galvanized garbage can with aeration holes drilled in the lid and top half of the sides. She adds waste daily, merely lifting the lid and dropping the waste onto the surface of the bedding. Occasionally, a double-handful of peat is thrown on top of the mass. When the lid is lifted, masses of worms can be seen feeding on the recently deposited waste. Odor has not been a problem in this system, the worms have reproduced greatly, and the end product appears to be well-converted vermicompost.

Should you grind the garbage?

No. Eventually, any soft food waste will break down to become vermicompost, even citrus and melon rinds. I have mentioned that I do pulverize egg shells with a rolling pin to reduce their size. There is no question that worms can eat ground food waste more readily than large particles of food. A worm's mouth is tiny, and it has no teeth to break down food particles. However, part of my rationale for using worms inside the home to process food waste is to reduce dependence upon technology. The energy required to grind garbage, dilute it with water, and flush it down the drain, as well as for processing it at the wastewater treatment plant can be better used elsewhere. For me, to regularly grind garbage before feeding it to worms is inconsistent with why I use worms in the first place.

Overloading the system

A common, and appropriate question is, "Can I put too much garbage in the worm bin?" Yes. You may have a greater than normal quantity of food waste during holidays, or during canning season. If you deposit all of it in your worm bin, you may find that you have overloaded the system. When this happens, it is more likely that anaerobic conditions will develop, causing odor. If you can leave it long enough without adding any fresh waste, the problem will usually correct itself. This does present you with the problem of how to dispose of your normal quantity of food waste during the interim.

A possible approach to the "overload" times is to set up a separate container with fresh bedding and use a half-bucket of vermicompost from your original bin to *inoculate* this new container with worms and microorganisms. This bin could be maintained minimally, feeding the worms only on the occasions when your week's garbage far exceeds the amount for which your main bin was constructed.

I have used for a "worm bin annex" an old leaky galvanized washtub, kept outside near the garage. During canning season the grape pulp, corn cobs, corn husks, bean cuttings, and other fall harvest residues went into this container. It got soggy when it rained, and the worms got huge from all the food and moisture. We brought it inside at about the time of the first frost. The worms kept working the material until there was no food left. After six to eight months, the only identifiable remains were a few corn cobs, squash seeds, tomato skins, and some undecomposed corn husks. The rest was an excellent batch of worm castings and a very few hardy, undernourished worms.

In other words, given enough time, practically any amount of organic material will eventually break down and decompose in a worm culture. When you want to add fresh material every week, as you do in a system being used to dispose of kitchen waste, there are limits to what is reasonable to add at one time. Your nose is probably the best guide as to when that limit has been exceeded.

Do I have to stay home and take care of my worms?

One favorable aspect of having "worms as pets" is that you can go away without having to make arrangements with the vet or a neighbor. You can go away for a weekend, a week, even two weeks, and not worry about your worms. However, if you plan to go for a month or more, or plan to turn off the heat during a winter vacation, you should probably board them out while you are gone.

10.
How do I
take care of my worms?

Tender loving care for worms means basically to provide them with the proper environment, check them occasionally, and leave them alone. The less you disturb them, the better off the worms are, although you will want to make some observations of what goes on in the box.

Once your worm bin is set up with bedding of the proper moisture content and a sheet of plastic lying loosely on top to retain that moisture, daily care is unnecessary. Burial of garbage, whether it is done weekly or more often, consists merely of pushing bedding aside to create a large enough pocket to contain the garbage, depositing the garbage, and covering it with an inch or so of bedding. Train yourself to make a few observations at these times. Does the bedding seem to be drying around the edges? Where are the worms congregating? To find out, you will have to push bedding aside in areas where you have deposited garbage. You can use your hands to do this, or you may prefer to use a trowel or a small hand tool similar to what I call my "worm fork." The worm fork is less likely to injure worms than a trowel.

Sometimes you will see masses of worms feeding around something that especially appeals to them. For curiosity's sake, you might want to note their preference. My worms, for example, love watermelon rind. I place the rind, flesh side down, on the surface of the bedding. Within the next two days, I will find masses of worms of all ages underneath the rind. Within three weeks, all that remains is the very outer part of the rind, looking a lot like a sheet of paper. The same is true for cantaloupe, pumpkin, and squash.

There are many other things you can observe. Do older worms prefer different food than younger worms? When do you first find cocoons? Are they deposited on top or throughout the bedding? Are any worms mating? Do you see differences in the degree to which the clitellum is swollen?

The preceeding questions give you an idea of the rich learning experience a home vermicomposting system can provide. Children are fascinated by worms. Many will find the system is an ideal science project. Even a three-year-old was able to understand the concept of feeding garbage to worms. She asked her mother, "Mommy, do I throw this in the garbage can, or do I feed it to the worms?"

Record keeping

I mentioned previously that I keep records of my worm bin. In fact, my records from the past ten years have provided much of the information in this book. Some of you will find that you want to keep records also, although for others, this would be a distasteful chore. If you decide to keep records, it will help if you have a utility scale to weigh the garbage, and a thermometer to determine bedding temperature. I currently use a data sheet similar to that appearing in the appendix.

Harvesting worms and changing the bedding

It may take about six weeks before you begin to see noticeable changes in the bedding. It will get darker, and you will be able to identify individual castings. Although food waste is being added regularly, the bedding volume will slowly decrease. As more of the bedding and garbage is converted to earthworm castings, extensive decomposition and composting by other organisms in the bin takes place. As mentioned earlier, as the proportion of castings increases, the quality of environment for the worms decreases. There will come a time when so much of the bedding in the box becomes castings that the worm population will suffer. Because each system is different—depending upon bedding used, quantity of worms, types of garbage fed to them, and temperature and moisture conditions—it is not possible to predict precisely when you must deal with changing the environment of your worms by getting them away from their castings and preparing fresh bedding.

Your particular goals, described in Chap. 1 in terms of whether they required high, low, or medium levels of maintenance, will help decide this. That is, to harvest extra worms for fishing, you will have to change bedding more frequently. Plan on doing this every two to three months, and figure that it is a high maintenance system.

If you don't want to harvest worms from partially decomposed garbage and bedding, but do want high quality vermicompost almost fully converted to worm castings, you will have to accept the trade-off of losing your worm population. In Northern systems, for example, you might bury garbage in your bin for the four winter months, and then let it sit unattended for another three to four months. By July, you will find a bin full of fine, black worm castings, but there will be very few worms remaining—perhaps not more than a dozen. These fine castings can be used as top dressing on your house plants and in your garden for a late shot of nutrients. This was referred to earlier as the "lazy person's" technique for maintaining a worm bin. I've done it, and it does work. When I'm using this system though, I have to compost food waste in outdoor compost piles during spring, summer, and fall.

High and medium maintenance systems require that you harvest the worms, or at least give them the opportunity to move into fresh bedding. For the high maintenance system, plan to do this every two to three months while medium maintenance can go to about four months. It will take two to three hours the first time you do this, but goes faster when you have some experience. If you have curious friends or family to help, it may go even faster. The illustrator of this book finds it therapeutic to sort worms. Her illustrations come from years of first-hand experience.

Dump and hand sort.

Needed:

- Very large sheet of heavy plastic to lay on the floor, table, or outside on the ground
- Goose-neck or similar lamp with 100 watt bulb, if you work inside
- Plastic dish pan or other container for worms
- Plastic or metal garbage can, corrugated carton, or heavy-duty plastic bag for vermicompost
- Fresh bedding

Spread the plastic sheet on the ground or table, and dump the entire contents of the worm bin on the plastic. Make about nine cone-shaped piles. You should see worms all over the place. If the light is bright enough, they quickly move away from it towards the center of each pile of vermicompost. If you are impatient, you can start hand-picking worms from this point on, gently scraping compost from the top of each pile, putting worms into the worm container as you find them.

You may prefer to leave the piles alone five to ten minutes. When you return you won't see any worms. Gently remove the outer surface of each pile. As you do so, worms on the newly exposed surface will again react to the light and retreat towards the interior. By following this procedure one pile at a time, you will find that when you return to the first pile, the worms will have disappeared again, and you can repeat the procedure.

Eventually, the worms will have aggregated in a mass at the bottom of each pile. Place these in your container, and remove vermicompost that collects on top of the worms. You will be amazed at how you are able to get a mass of "pure worms" using this technique. You need to remove all vermicompost from this batch only if you are going to weigh the worms. During this process, you or someone else should be making up fresh bedding and restocking the worm bin. When the bedding is ready and the worms are sorted and weighed, add them to the top of the bin as you did originally and you are ready for another cycle.

HARVESTING TECHNIQUES:
DUMP AND HAND SORT.

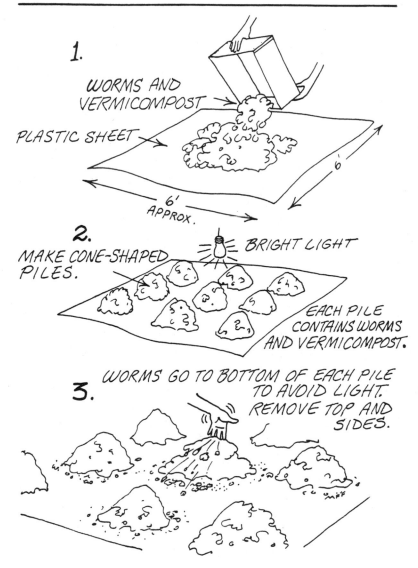

1. WORMS AND VERMICOMPOST

PLASTIC SHEET

6'

6' APPROX.

2. MAKE CONE-SHAPED PILES.

BRIGHT LIGHT

EACH PILE CONTAINS WORMS AND VERMICOMPOST.

3. WORMS GO TO BOTTOM OF EACH PILE TO AVOID LIGHT. REMOVE TOP AND SIDES.

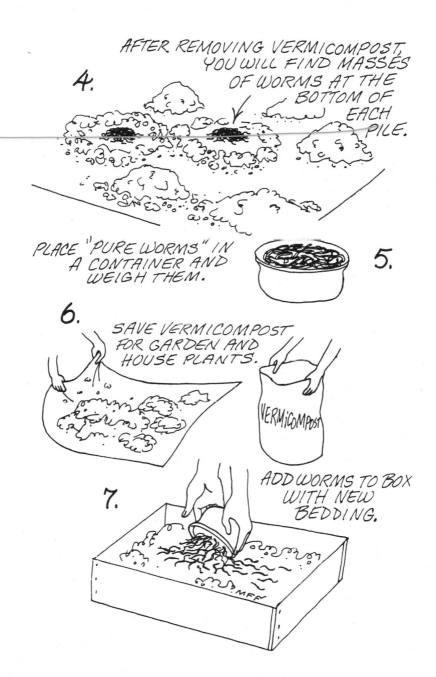

4. AFTER REMOVING VERMICOMPOST, YOU WILL FIND MASSES OF WORMS AT THE BOTTOM OF EACH PILE.

PLACE "PURE WORMS" IN A CONTAINER AND WEIGH THEM. **5.**

6. SAVE VERMICOMPOST FOR GARDEN AND HOUSE PLANTS.

VERMICOMPOST

7. ADD WORMS TO BOX WITH NEW BEDDING.

Vermicompost from this sorting process will vary in consistency, depending upon how long the bin has been going, how much and what kind of garbage was buried, and how much decomposition has occurred. Some of the most recently buried food waste can be put right back in the fresh bedding. The rest will continue to compost in a plastic bag, garbage can, or corrugated carton. Eventually, as it dries, it can be used for the garden or your house plants, as described more fully in Chap. 13.

A large number of cocoons and baby worms will be present in this vermicompost. If you wish, you can save many of them by letting the vermicompost sit for about three weeks. Then attract them to a long, narrow strip of food. (This may be one occasion where use of a blender is appropriate.) Make a slurry of garbage, perhaps with some oatmeal, cornmeal, or other grain mash in it. With your fingers or a trowel, make a groove down the center of the vermicompost, and fill this groove with the slurry. In a couple of days, you should be able to remove concentrated batches of young worms from underneath this narrow strip. Repeat two or three times to obtain new hatchlings as they come along. You can add these new worms to your regular bin.

Let the worms do the sorting. If you don't want to deal with the "Dump and Sort" method described above, there are some ways to avoid that process if you don't care to know the weight of your worms. When the bedding has diminished to the extent that it is not deep enough to make a hole to bury fresh garbage, it is time to add fresh bedding. Prepare about one-half the original quantity of fresh bedding. Pull all of the vermicompost in your bin over to one side, and add the new bedding on the other side. Bury your garbage in the new bedding and let the worms find their way to it. It is helpful to replace the plastic sheet only on the side with the fresh bedding to permit the other side to dry out more rapidly.

Every two to three months, you can remove the vermicompost, replace it with more fresh bedding, and keep going back and forth from one side to the other in this manner. The vermicompost you remove will still have some worms in it, but enough should have migrated to the new bedding so that you

HARVESTING TECHNIQUE:
LET THE WORMS DO THE SORTING.

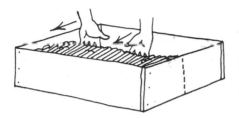

1.

PULL VERMICOMPOST AND WORMS TO ONE SIDE OF THE BOX.

2.

ADD NEW BEDDING TO VACANT SIDE.

3.

BURY GARBAGE IN NEW BEDDING.

WORMS MOVE TO NEW BEDDING IN SEARCH OF FOOD.

4. BLACK PLASTIC THIS SIDE ONLY

REMOVE VERMICOMPOST IN 2~3 MONTHS AND REPLACE WITH NEW BEDDING.

needn't worry about harvesting the few that remain.

Divide and dump. Still another method for harvesting worms is the "Divide and Dump" technique. You can simply remove about two-thirds of your vermicompost and dump it directly on your garden. Add fresh bedding to the vermicompost that is still in the box. Enough worms and cocoons usually remain to populate the system for another cycle.

The maintenance system you use will depend upon your preference, your lifestyle, and perhaps your schedule at the time. You may find yourself using all of these systems at various times. At any rate, maintaining your home vermicomposting system can be a flexible process and is really very simple.

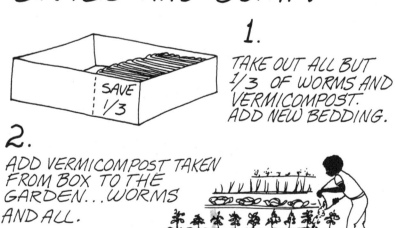

HARVESTING TECHNIQUE:
DIVIDE AND DUMP.

1.
TAKE OUT ALL BUT 1/3 OF WORMS AND VERMICOMPOST. ADD NEW BEDDING.

2.
ADD VERMICOMPOST TAKEN FROM BOX TO THE GARDEN...WORMS AND ALL.

SAVE 1/3

11.
What are the most commonly asked questions about worms?

Are you one of the thousands of people who have mixed reactions towards worms? Do you feel revulsion towards these moist, wriggly creatures at the same time that you are fascinated by them? Are you somewhat curious, but don't want to learn too much about them? Then this chapter is for you. If reading this far has merely whetted your appetite to learn more about earthworms, however, refer to the publications listed under selected references for books with more detailed information.

Can a worm see?

Contrary to the popular cartoon image of worms, they have no eyes so they cannot see. They are, however, sensitive to light, particularly at their front end. If a worm has been in the dark and is then exposed to bright light, it will quickly try to move away from the light. A nightcrawler, for example, will immediately retract into its burrow if you shine a flashlight on it some wet spring night.

The sensory cells in a worm's skin are less sensitive to red light than to light of mixed wavelengths. If you want to observe worms, you can use this fact to your advantage by placing red cellophane or an amber bread wrapper over your light source. You can make more observations of earthworm behavior in a photographic darkroom using a red safelight. Your eyes will adapt to the low light levels, and the worm will move more naturally than it does under bright light.

Where is the worm's mouth?

A worm's front and back ends are more technically known as anterior and posterior. The mouth is in the first anterior segment. A small, sensitive pad of flesh, called the *prostomium*, protrudes above the mouth. When the anterior end of the worm contracts, the prostomium is likely to plug the entrance to the mouth. When the worm is foraging for food, the prostomium stretches out, sensing suitable particles for the worm to ingest.

Does a worm have teeth?

No. The mouth and pharynx are highly muscular, but they do not contain teeth.

How does a worm grind its food?

Because worms have no teeth, they have little capacity to grind their food. They are limited to food that is small enough to draw into their mouth. Usually this food is softened by moisture or by bacterial action. Much of it undoubtedly is bacteria, protozoa, and fungi, which break down the organic material ingested. Worms do have a muscular gizzard, which functions similarly to that of birds. Small grains of sand and mineral particles lodge in this gizzard. The muscular contractions compress these hard materials against each other and the food, mix it with some fluid, and grind it into smaller particles. One reason for mixing a handful of topsoil or lime into worm bedding is to provide worms with small, hard particles for their gizzard.

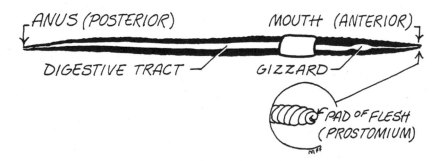

What happens to food once it leaves the gizzard?

The ground up food enters the worm's intestine, which secretes digestive enzymes. The enzymes chemically break down the molecules, which then pass through the intestinal wall to be absorbed into the bloodstream and carried where needed. Undigested material, including soil, bacteria, and plant residues, is passed out of the worm through the anus as a worm casting.

If a worm is cut in half, will both parts grow back?

Worms do have a remarkable capacity to regenerate lost or injured parts, but this capacity is limited. Depending upon where the worm was cut, the anterior end can grow a new tail. The tail, however, cannot regenerate a new head. The capacity to regenerate new tissue is a form of reproduction among some animal forms, but not among earthworms. On rare occasions, you may find a worm with two tails. This condition can be caused by injury to the worm in the posterior end, which results in growth of a new tail adjacent to the original tail.

The cutting-a-worm-in-half myth.

Do worms die in the box?

Worms undoubtedly die in any home worm bin, but if your box is properly maintained, you will rarely see a dead worm. Their bodies quickly decompose and are cleaned up by the other organisms in the box, leaving few dead worms you can recognize.

If large quantities of worms seem to be dying, you should attempt to determine the cause and correct the problem. Is it too hot? Are toxic gases building up in the bedding, which cause the worms to surface to get away? Did you stress the worms by adding too much salty food or acid-producing food? You'll need to make some educated guesses about what the problem is, and try to correct it. Sometimes, adding fresh bedding to a portion of the box is enough to correct the situation by providing a safe environment towards which the worms can crawl.

How long does a worm live?

Most worms probably live and die within the same year. Especially in the field, most species are exposed to hazards such as dryness, too cold or hot weather, lack of food, or predators. In culture, the individuals of *Eisenia foetida* have been kept as long as four and a half years, and some *Lumbricus terrestris* have lived even longer.

Do worms need air?

Worms require gaseous oxygen from the air. The oxygen diffuses across the moist tissue of their skin, from the region of greater concentration of oxygen (the air) to that of lower concentration (inside the worm). When water has been sufficiently aerated, worms have been known to live under water for a considerable length of time.

Carbon dioxide produced by the bodily processes of the worm also diffuses through its moist skin. Also moving from higher concentration to lesser concentration, carbon dioxide moves from inside the worm's body out into the surrounding bedding. A constant supply of fresh air throughout the bedding helps this desirable exchange of gases to take place.

12.
What are some
other critters in my worm bin?

Once your home vermicomposting system has been established for awhile, you will begin to find creatures other than earthworms present. This is a normal situation, but could be alarming if you were brought up to think that "all bugs are bad bugs." Most of them are, in fact, good bugs. They play important roles in breaking down organic materials to simpler forms that can then be reassembled into other kinds of living tissue. This whole array of decomposer organisms constitute the true recyclers. You could spend a lifetime studying the various creatures in a worm bin trying to determine who eats whom, and under what conditions.

A scientist who has spent years studying the complex interrelationships among organisms found in compost piles and decaying litter is Dr. Dan Dindal. It was he who suggested, in Chap. 9, grinding meat, and mixing it with a carbon source. Dr. Dindal developed the drawing in Figure 3 to illustrate what many of these organisms look like and how they relate to each other.

Referring to Fig. 3 you can see that organic residue is eaten by first level (1°) consumers, such as molds and bacteria. Earthworms, beetle mites, sowbugs, enchytraeids and flies also consume waste directly. First level consumers are eaten by second level (2°) consumers, such as springtails, mold mites, feather-winged beetles, protozoa and rotifers. Third level consumers are flesh-eaters, or predators, which eat 1° and 2° consumers. Predators in a compost pile, or in your worm bin, might include centipedes, rove beetles, ants, and predatory mites.

You won't be able to see many of the organisms pictured because they are microscopic (bacteria, protozoa, nematodes,

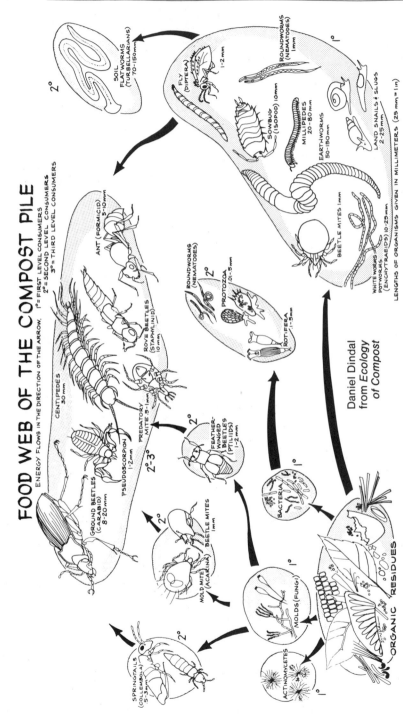

FOOD WEB OF THE COMPOST PILE

ENERGY FLOWS IN THE DIRECTION OF THE ARROW.   1° = FIRST LEVEL CONSUMERS
2° = SECOND LEVEL CONSUMERS
3° = THIRD LEVEL CONSUMERS

SOIL
FLATWORMS
(TURBELLARIANS)
70-150mm   2°

FLY
(DIPTERA)
1-2 mm

ROUNDWORMS
(NEMATODES)
1mm   1°

SOWBUG
(ISOPOD) 10mm

MILLIPEDES
20-80mm

LAND SNAILS & SLUGS
2-25mm

EARTHWORMS
50-150mm

BEETLE MITES 1mm

WHITE WORMS
POTWORMS
(ENCHYTRAEIDS) 10-25mm

LENGTHS OF ORGANISMS GIVEN IN MILLIMETERS (25 mm = 1 in)

CENTIPEDES
30 mm

ANT (FORMICID)
5-10mm

ROVE BEETLES
(STAPHYLINID)
10mm

ROUNDWORMS
(NEMATODES)   2°

PROTOZOA
.01-.5mm

ROTIFERA
.1-.5mm

GROUND BEETLES
(CARABID)
8-20 mm

PSEUDOSCORPION
1-2mm

PREDATORY
MITE .5-1mm   2°-3°

FEATHER-
WINGED
BEETLES
(PTILIDS)
1-2 mm   2°

MOLD MITE
(ACARINA)
2°

BEETLE MITES
1mm

BACTERIA   1°

MOLDS (FUNGI)   1°

ORGANIC RESIDUES

SPRINGTAILS
(COLLEMBOLA)
.5-5 mm   2°

ACTINOMYCETES   1°

Daniel Dindal
from *Ecology
of Compost*

Figure 3. Organisms commonly found in compost. Energy flows from organism to organism as one is eaten by the other in a natural recycling system. Snails, beetles, millipedes, centipedes, and ants are less likely to find their way to worm bins set up with shredded corrugated and paper beddings.

and rotifers). Others, such as the springtails and mites, are so small that you will probably need a hand lens to get a better look. Brief descriptions of the more common "critters" follow.

Enchytraeids

Known commonly as white worms, or pot worms, enchytraeids are small (1/4 to 1 inch long), white, segmented worms. They are so tiny they look as if they might be newly hatched redworms. However, redworms have red blood; even newly hatched redworms are reddish. Although related to the larger earthworms, enchytraeids do not have a hemoglobin-based blood, and remain white throughout their lifetime.

Some worm growers incorrectly call enchytraeids, "nematodes," and feel that they should try to get rid of them. Nematodes are undoubtedly present in large quantities in worm bins, but you would not be likely to see them without a microscope. Some commercial worm growers are concerned that enchytraeids will compete with redworms for feed, and may attempt to control their numbers. Since the purpose of having a home vermicomposting system is to get rid of food waste, the presence of an organism that helps to do the job is an asset, not a detriment. My position concerning enchytraeids is, "if they be, let them be."

Springtails

In your worm bin, you may see a sprinkling of hundreds of tiny (1/16 inch) white creatures against the dark background of the decomposing bedding. When you reach your finger toward them, some spring away in all directions. Springtails are primitive insects with a pointed prong extending forward underneath their abdomen from the rear. By quickly extending this "spring," they jump all over the place. Other members of this group, which scientists call *Collembola*, do not have the springing tail. Collembola feed on molds and decaying matter and are important producers of humus.

Springtails are not only numerous, they are diverse, with over 1200 species described. They live in all layers and types of soils, and are considered to be among the most important soil organisms.

Isopods (sowbugs and pill bugs)

Isopods are easy to identify because the series of flattened plates on their bodies makes them look like tiny armadillos. If one of these grey, or brown, half-inch long creatures rolls up in a ball, it is commonly called a pill bug. Its scientific name is *Armadillidium vulgare*. Sowbugs are related to pill bugs, but don't roll up in a ball.

If you have used manure as a part of the bedding in your worm bin, you are almost certain to have a few isopods grazing over the surface. They won't do the worms any harm since they are vegetation and leaf litter eaters.

Centipedes

Centipedes are just about the only critters that I kill on sight in a worm bin. You'll probably never have very many of them, but they are predators, and they do occasionally kill worms. Centipedes move quickly on their many legs. You can tell a centipede from a millipede (thousand-legged) by looking carefully at the attachment of its legs to its body. If it has only one pair of legs per segment, it is a centipede. Millipedes have two pairs of legs on each segment.

Millipedes

You may find a few millipedes in your worm bin, especially if you used manure, leaf mold, or compost as part of your bedding. They are vegetarians and won't kill your worms. In fact, they are very helpful and contribute more to breaking down organic matter than generally realized. After leaves have been softened by water and bacteria, millipedes eat holes in them, helping springtails, mites, and other litter dwellers skeletonize them so that only the ribs remain. I wouldn't even consider killing a millipede.

Mites

You will undoubtedly have many, many mites in your worm bin. Like the springtails, mites are so small it is difficult to see them, except as minute dots moving across the surface of the bedding. Mites have eight legs and a round body. Some eat plant materials, such as mold, decaying wood, and soft tissues of leaves. Others consume the excrement of other organisms. Beetle mites don't travel very far on their own, but they travel as stowaways on dung beetles, which transport them from one dung heap to another.

One kind of mite, known as the earthworm mite, can be a problem in worm beds. This mite is brown to reddish, and can achieve such high numbers that the worms may refuse to feed. They are more likely to be present in very wet beds, and may concentrate on one or another kind of food, completely covering the surface. If this happens, remove and burn the mite infested food, or put it out in the sun to kill the mites. Bait others in the same way, or by placing a piece of bread on the bedding and removing it when the mites concentrate on its underside.

To create conditions that aren't so favorable for earthworm mites, leave the cover off your bed for a few days to reduce bedding moisture. I have seen only one or two bad infestations of red mites in ten years of using worms to eat my garbage, so I don't consider them a serious problem.

Fruit flies

If someone were to ask me, "What is the most annoying problem you have encountered in having a worm bin in your home?", I would have to answer, "Fruit flies—not odor, not maintenance, not worm crawls, but fruit flies." Not every box has them, and not every box that has them has them all the time, but when they are present, they are a nuisance.

Fruit flies are not dangerous. They don't bite, itch, or buzz. When they land in your orange juice, beer or champagne, however, it's a bit much. Even the most tolerant guests figure that maybe you've gone too far in this "ecology thing."

To date, I have found no sure-fire method to get rid of fruit flies. I prefer not to contaminate my home, the air I breathe, my worms, and my garden with toxic pesticides, so I am not willing to use products that leave harmful residues to eliminate fruit flies. There are some things you can do, however, to try to control them.

Traps. Usually, fruit flies are most numerous in and around the worm bin. They undoubtedly come in on fruit peels and rinds, or are attracted by them in late summer or early fall. Bin conditions are favorable for their reproduction, and they are very prolific. A simple trap will attract large numbers of them. It doesn't totally get rid of them, but it does keep them somewhat under control.

Materials required:
  jar   plastic bag   beer   rubber band

To make the trap, pour a half-cup or so of beer into the jar. Place the plastic bag over the mouth of the jar with one corner reaching down into the jar. Poke a small (no more than 1/4 inch diameter) hole in the corner of the bag with a pencil. Secure the bag around the rim with the rubber band. Fruit flies will be attracted by the fermenting beer, find their way through the tiny hole in the bottom of the funnel, and not be able to find their way out. Voila! You'll catch hundreds of them. They seem to die when beer is used as the attractant.

FRUIT FLY TRAP
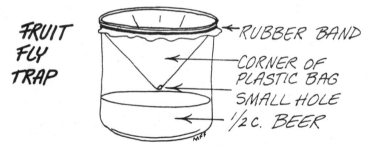

Other foods, such as orange juice, overripe banana with water, grapefruit, and sourdough, attract them too. When you use these, however, the flies lay eggs that hatch. The larvae form pupae, which then hatch to become adults and you end up with more fruit flies, even if they *are* trapped in the jar. One way to curtail this unwanted population explosion is to change the culture medium (banana, etc.) every week or so. I have run hot water into the jar to kill the adults and larvae, then set it up again.* With beer, I avoid that weekly chore.

Vacuum. Although this method deserves a minus from the standpoint of environmental soundness because it requires electricity, if you get desperate, suck them up with a vacuum cleaner. When I lift the plastic cover on my box, lots of fruit

---

*Many gardening books suggest that you flip potato beetles, tomato hornworms, slugs and other hand-collectable pests into a container of kerosene to kill them. I don't bother with kerosene because hot water does just as well. It rapidly coagulates the protein in the insects, killing them immediately. If the water isn't very hot, add some detergent to the water, and that also kills the pests. These carcasses can go into your worm bin.

flies fly up and land on the basement ceiling. Then I just inhale them with the vacuum cleaner. Although not a complete control, this method does help to cut down the numbers.

Other suggestions. Diatomaceous earth sprinkled on the surface of the bedding hasn't worked. Yellow plastic discs covered with mineral oil or honey to attract and hold fruit flies haven't worked either, although use of a stickier adhesive might work.

Suggestions from friends who have compost toilets have included getting a cow patty from a field and putting it into the box. Beetles in the cow manure are said to feed on the fruit fly larvae and serve as an effective biological control. Another method described by the compost toilet people is to incorporate a solar cooking phase to treat all garbage before it is placed in the compost unit. They suggest heating it to the point where fruit flies, eggs, larvae, and pupae will be killed before putting garbage into the bin. Such preventive maintenance might work.

Ants

I have never had a problem with ants in my vermicomposting bin, but then I don't have a problem with ants in my home, either. In milder climates, though, ants could be a problem for which controls must be sought.

Some worm growers report that certain baits and insecticidal sprays are effective, but I would try physical barriers first. For example, I would set the legs of my worm bin in coffee cans with mineral oil in the bottom. The ants would get trapped in the oil and would not be able to enter the bin. Or, I might try dabbing Vaseline on a piece of cotton and making a continuous one-inch swath around the top of my worm bin. I don't know if it would work, but I would try it before I would resort to a commercial ant bait that might use arsenic as its primary killing agent.

Disease organisms

A question sometimes asked is, "Can you get viruses, germs, or diseases from your worm bin?" That's not a simple question to answer. I have already discussed the potential for transmitting toxoplasmosis if cats are allowed to use a worm bin as a litter box. The organism for this disease is known to pass intact through the digestive tract of an earthworm. If your cat is harboring the organism, it can pass into the cat's feces. The more you are exposed to the places the feces are deposited, the more likely the organism could enter your body. Of course, this could happen without a worm bin by merely changing the cat litter box.

For similar reasons I discourage people from even considering using a worm bin similar to that described in this book to handle human manure. Pathogens (disease producing organisms) can be transmitted in human manure. Our complex and increasingly expensive wastewater treatment facilities are designed to reduce or eliminate the possibility that these organisms will reach our soils and water supply. Although high-temperature composting has been shown to be effective in killing pathogens, home vermicomposting systems do not generate the high temperatures characteristic of well-constructed, large-mass compost heaps.

Some research has been done that suggests that passage through an earthworm's gut can reduce the number of pathogents present in sewage sludges. This is preliminary work, and more research needs to be done in this area. Until solid data are available, caution against using human waste is in order.

One further caution—if you are overly sensitive, or allergic to fungi and mold spores, you probably won't be able to have a worm bin in your home. Molds can and do develop as a natural sequence in the composting process. You may have to carry out your vermicomposting activities outdoors, perhaps with someone else doing the maintenance required. Another possibility would be to keep the acidity of the bin within a pH range of six to eight which is outside the optimal range of fungi (pH 4 to 6).

In summary, you are likely to find many organisms other than earthworms in your worm bin. In truth, the system won't work if they aren't present. Your worm culture is not a monoculture, but a diverse community of micro and macro-organisms that are interdependent. No one species can possibly overtake all the other species present. They serve as food for each other, they clean up each other's debris, they convert materials to forms that others can utilize, and they control each other's populations.

For us to arbitrarily decide who should live and who should die in this complex system is a bit presumptuous. Although some controls are suggested, this chapter's major purpose is to provide a better idea of what you can expect to find. That way you won't be alarmed, and you may even decide that you want to learn more about those critters that you always used to squash when you found them.

13.
How do plants benefit from a worm bin? Completing the circle

From the beginning I have tried to relate having worms eat your garbage to having healthier plants. This happens when you use the vermicompost from your worm bin on your house plants and gardens. What is the nature of this rich humus, and how should you use it?

It helps to remember the distinction between vermicompost and worm castings. Worm castings have moved through the digestive tract of a worm; vermicompost is a mixture of worm castings, organic material, and bedding in varying stages of decomposition, plus the living earthworms, cocoons, and other organisms present.

If you choose a low maintenance system, a large proportion of your vermicompost will be worm castings. A worm casting (also known as worm cast or vermicast) is a biologically active mound containing thousands of bacteria, enzymes, and remnants of plant materials and animal manures that were not digested by the earthworm. The composting process continues after a worm casting has been deposited. In fact, the bacterial population of a cast is much greater than the bacterial population of either ingested soil, or the earthworm's gut.

An important component of this dark mass is humus. Humus is a complicated material formed during the breakdown of organic matter. One of its components, humic acid, provides many binding sites for plant nutrients, such as calcium, iron, potassium, sulfur and phosphorus. These nutrients are stored in the humic acid molecule in a form readily available to plants, and are released when the plants require them. Humus increases the aggregation of soil particles which, in turn, enhances permeability of the soil to water and air. It also buffers the soil, reducing the detrimental effects of excessively acid or aklaline soils. Humus has also been shown to stimulate plant growth and to exert a beneficial control on plant pathogens, harmful fungi, nematodes, and harmful bacteria. One of the basic tenets of gardening organically is to carry out procedures that increase the humus component of the soil; earthworm activity certainly does this.

How to use vermicompost

You will have several buckets full of vermicompost from your worm bin. Use it selectively and sparingly. Vermicompost is loaded with humus, worm castings, and decomposing matter. Plant nutrients will be present both in stored and immediately available forms. Vermicompost also helps to hold moisture in the soil, which is an added advantage during dry periods.

Seed beds. Vermicompost will not burn your plants as some commercial fertilizers do, but since your supply will be limited, use it where it will do the most good. One method is to prepare your seed row with a hoe, making a shallow, narrow trench. Sprinkle vermicompost into the seed row. In this way, the new seeds will have the vermicompost as a rich source of nutrients soon after they germinate and during early stages of their growth.

SPRINKLE VERMICOMPOST INTO SEED ROW.

TRANSPLANTED SEEDLING

SOIL

VERMICOMPOST

Transplants. For transplanting such favorites as cabbage, broccoli and tomatoes, which are usually set out in the garden as young plants, throw a handful of vermicompost in the bottom of each hole you dig for a plant. Don't worry if worms or cocoons are present in the vermicompost. While the worms are alive, they will work and aerate the soil, produce castings, add nitrogen from their mucus, and do all the other good things that worms do for the soil. But don't expect your redworms to thrive in your garden. They are not normally a soil dwelling worm, and require large amounts of organic material to live. If you were to add large quantities of manure, leaves, or other organic material, you might get a few *Eisenia foetida* to live, but most will probably die. When they do, their bodies will add needed nitrogen to the soil, so all is not lost! Hopefully, your gardening techniques will improve the organic matter concentrations in your graden so that the soil dwelling species of earthworms will be fruitful and multiply.

Top dressing. Most of your supply of vermicompost from your winter's production will be used during spring planting. Any remaining can be used later in the season as top or side dressing. At this time you won't want to disturb the root systems, but it is a simple matter to sprinkle vermicompost around the base and drip line of your plants, giving them an additional supply of nutrients, providing organic matter, and enabling the mid-season plants to benefit from vermicompost's water-holding capacity.

DRIP LINE

TOP DRESS WITH VERMICOMPOST AS PLANT GROWS.

How to use worm castings

After several months of low maintenance technique, the contents of the worm bin will be a dark, earthy-smelling, crumbly material. Since little food is left in this material for earthworms, very few will be present. Populations of active microorganisms will also have dwindled; those present will be in a dormant state awaiting reactivation in a suitable environment of new food and moisture.

Except for some large chunks, most of this material is

worm castings. Worm castings differ from vermicompost in being more homogeneous, with few chunks of recognizable bedding or food waste. When dried and screened, castings look so much like plain, black topsoil that you may be surprised to recall that little or no soil went into the original bedding.

While some drying of worm castings is desirable, it is best not to let them dry to the point where they become powdery, for it then becomes difficult to wet them down. Worm castings with about a 25 to 35% moisture content have a good, crumbly texture and earthy smell, and are just about right to use on your plants.

Although pure worm castings provide many nutrients for plants in a form the plants can use as needed, some precautions should be taken in their use. The organic material present in food waste is likely to have been broken down to a greater extent in worm castings than in vermicompost. More carbon will have been oxidized and given off as carbon dioxide, leaving phosphorus, potassium, calcium, magnesium and other elements to combine to form salts. High concentrations of salts can inhibit plant growth. Worm castings, which may have high concentrations of salts, should be diluted with other potting materials so that plants gain the advantage of the nutrients present without suffering from the possible high concentration of salts.

An intriguing experiment, which seemed to verify the need to dilute pure castings, was conducted by a horticulturist at the Kalamazoo (Michigan) Nature Center. Three sets of African violet plants were potted, each set in a different medium. (See Fig. 4) The plants on the left, labeled PS, had 100% potting soil, C (right) had 100% worm castings, and C-P-MP (center) contained equal amounts of worm castings, perlite, and Michigan peat. Although difficult to see in this black and white photograph, a comparison of the plants grown in potting soil with those grown in pure worm castings shows the castings-grown plants to be healthier looking. Those grown in the potting soil were beginning to show chlorosis, or yellowing of some leaves, a sign of possible nutrient deficiency.

The center row of plants, grown with worm castings

Figure 4. Response of African violets to potting mixtures with and without worm castings. PS (1) = 100% potting soil, C (r) = 100% worm castings, C-P-MP (c) = 1/3 worm castings, 1/3 perlite, and 1/3 Michigan peat. Those with a diluted mixture of worm castings were larger, greener, and showed more vigorous growth. The experiment was conducted by horticulturist Cheryl Lyon at the Kalamazoo Nature Center.

diluted with perlite and peat, are distinctly more vigorous than either of the other two sets of plants. Leaves are larger, greener, and more robust. A likely interpretation of this experiment is that, although the 100% castings provided more nutrients for the young plants than the potting soil, salt concentrations in the castings may have been great enough to inhibit their growth. The center row of plants had the benefit of nutrients from the castings, but was not inhibited by too high a concentration of salts, since the concentration had been reduced by dilution with the perlite and peat.

Of course, other interpretations are possible. For example, the center row of plants may have had an advantage contributed by the water-holding capacity of the peat, plus the increased lightness of the potting mixture due to the perlite. Further experiments could discriminate between these possibilities, but from this preliminary work, it seems safe to say that some castings are better than no castings, and pure castings may not be as good as other possible mixes.

Much more research needs to be done to determine the true effects of worm castings on plant growth. Because the nutrient content of worm castings is directly dependent upon the types of organic materials the worms ate to produce those castings, castings themselves are highly variable. Plants also vary in their need for nutrients and many kinds of plants should be tested under many kinds of growing conditions.

Other sources of worm castings may eventually become available. Some work has been done using worms to process sewage sludge. There is a strong possibility that castings from municipal sewage sludge will contain higher concentrations of heavy metals than should be placed on soils in unlimited quantities. Will the presence of these heavy metals affect plant growth, or, if they are taken up by the plants, will they be a problem to us when we eat the plants? Although preliminary data are available on some of these important questions, there are many more questions than answers to date.

To sterilize or not to sterilize castings. Some people suggest "sterilizing" potting mixes and/or worm castings, prior to use in house plants and greenhouses, to kill organisms that could cause the plants trouble in a confined environment. The term "sterilizing" is being used loosely here, since sterilization literally means "the destruction of all living microorganisms, as pathogenic or saprophytic bacteria, vegetative forms, and spores."* Surgical instruments, for example, are sterilized in an autoclave under high temperature and pressure for a specified period of time. For our purposes, it would be more correct to say that potting soil is pasteurized; that is, it is exposed to a high temperature or poisonous gas for a long enough period of time to kill certain microorganisms, but not all. In any case, whether you prefer to call it sterilize or pasteurize, I don't recommend that you do either to worm castings.

Soil is a dynamic, living entity, and much of its value comes from the millions of microorganisms present. Chemical tests of worm castings often show fairly low amounts of nutrients present, yet plants grown in the "low testing" material still have higher yields than those grown with high concentrations of commercial fertilizers. Worm castings are reportedly being used in Japan on a very large scale in conjunction with commercial fertilizers in order to use lesser amounts of this increasingly expensive commodity.

One concern many people have about using worm castings directly on their house plants is, "Won't those little white worms and all those bugs I can see crawling around hurt my plants?" Probably not. The enchytraeids eat dead and decaying material, not living plants, and so do the mites and springtails that are likely to still be present when your vermicompost is almost all worm castings. The organisms that thrived in your worm box are not likely to be the kind that also attack living plants. If there are just a few, don't worry about them.

If there are a lot, and you have a true aversion to having visible critters in the worm castings you want to sprinkle under your plants, place your worm castings on a sheet of plastic outdoors in the sun, put another sheet of plastic on top, and let this "solar heater" warm things up a bit. Most of the white worms will move onto the plastic, and most of the mites and springtails will be killed from the heat. Collect your castings in a few hours and they will be ready to use in potting mixes, as top dressing, or in your garden.

*Random House Dictionary of the English Language, Unabridged Edition, 1969.

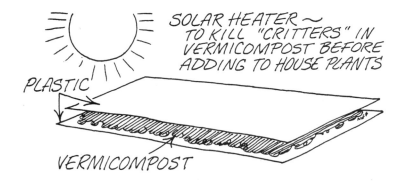

SOLAR HEATER ~
TO KILL "CRITTERS" IN
VERMICOMPOST BEFORE
ADDING TO HOUSE PLANTS
PLASTIC
VERMICOMPOST

Potting mixes. Worm castings may be mixed with various concentrations of potting materials, such as peat moss, sand, top soil, perlite, vermiculite, or leaf mold. One satisfactory mix is:

1/4 worm castings. . . . . for nutrients
1/4 peat moss. . . . . for moisture retention
1/4 perlite. . . . . for aeration
1/4 sand or garden soil. . . . . for body

Experiment with different mixes, and find the ones that suit you for your favorite plants.

Top dressing. Sprinkle worm castings about one-fourth inch deep on the surface of your potted plants and water as usual. Repeat every 45 to 60 days. If necessary, remove some of the soil above the roots so that you have room for the worm castings. Let an excess of water move through the soil occasionally to flush out accumulations of salts, particularly if you have hard water. And remember, don't use softened water on your plants; it will contain salts that could harm them.

In your garden. Sprinkle worm castings in the bottom of your seed row, or throw a handful of castings in the hole at the time you are transplanting. The adjacent soil will dilute excessive salt concentrations in the castings. It is perfectly natural for vegetable seeds present in the vermicompost to sprout. Simply pull these sprouts as you would pull weeds.

An added bonus: avocado plants!

Have you ever tried to germinate an avocado pit? Have you tried the trick with the three toothpicks, inserting them around the diameter of the pit, placing it on top of a jar of water, and keeping it watered for . . . well, months? Until you either got tired of it, or it finally did sprout?

Well, have we got a deal for you. Throw your pits in your worm bin, cover, and forget about them. That's all. In time — it may take months, but it'll happen — you will find a tap root coming out of the bottom, and a sprout coming out of the top. When this happens, transfer it to a pot. One winter nine out of ten avocado pits I tried this way germinated. I now have more avocado plants than I know what to do with — in the living room, on the front porch, on the side porch, in my office . . . . You, too, can be the first on your block to be a success with your avocado pits.

USE CASTINGS TO TOP DRESS HOUSE PLANTS.

REPEAT EVERY 45-60 DAYS.

TAKE OUT 1/4 INCH OF SOIL TO ALLOW ROOM FOR CASTINGS.

14.
What other ways can I recycle?

Once you have a home vermicomposting system, you will begin to see how recycling materials on location is extremely effective. I hope you'll even ask questions, such as, "Why am I continuing to spend good money for garbage pickup? Where does all that trash go, and why? Can't some of it be recycled, too? Aren't we going to run out of raw materials if we just keep throwing things away after using them?"

If you begin to ask yourself those questions, I hope you will spend some time and effort finding answers. Specific answers vary depending upon where you live, how progressive your community is, how responsible your public officials are, how environmentally conscious your businesses are, and how aware you and your fellow citizens are. It's hopeful that we seem to be in a time when individuals are trying to recover some control over their own destiny. Finding out how kitchen waste can be processed by worms is a good start, but there are other things you can do to take more control over the materials in your life. (A much more comprehensive discussion can be found in Jerome Goldstein's book *Recycling*.)

Reduce

Reduce the amount of materials that flow into and out of your home. Buy for quality and longevity when possible. Repair when you can, and, through your buying habits, encourage production of products that can be repaired. Refuse excess packaging by buying items that have less packaging. Encourage production of packaging that can be reused and recycled, rather than mixtures of plastics and paper, which are good for neither reuse nor recycling.

Reuse

Reuse what you can—plastic bags, jars, boxes. Plastic milk jugs can be used to store soft, non-chlorinated rainwater for watering your house plants. Nails can be pulled from scrap lumber, which can then be reused for building projects. Save your charcoal by pouring water over the coals when you have taken the food off your grill. Develop a conservation (some people might say "depression") mentality about the use of materials.

Recycle

Recycle newspapers, cans, glass, and aluminum whenever you can. The number of communities offering effective recycling programs is increasing every year. As the cost of obtaining virgin materials increases, incentives for recycling already-processed materials also increase, creating larger and more reliable markets for recycled materials. During the next decade, it should become easier to establish recycling programs. Since not only economic, but political forces, influence the direction these programs will take, you can help. Become involved in environmentally sound programs, not only in your own home, but in your community.

While the preceeding ideas are extremely general, they provide some basic priorities for environmentally conscious people. They are appropriate suggestions for moving away from our consumer orientation towards one of living gently on the earth. There are some other specific actions you can take to help keep waste disposal costs down and acknowledge your role as caretaker.

Recycle newspapers, junkmail, and paperboard packaging. It's not unusual to generate 20 to 30 shopping bags of newspapers, junkmail and paperboard packaging every three to four months. Eliminating contaminants, such as carbon paper, plastic, glassine envelope fractions, filament tape, and staples, I combine papers with friends and neighbors, and take them to our local recycling center where I am paid about a penny a pound. That pays for gas, with a few dollars to spare. This requires storage capacity (garage) and some effort, but saves recyclable paper from going to the dump and helps eliminate the necessity for weekly trash pickup. It also saves trees.

Flatten cans and reuse jars. I avoid paying the amount it would cost for trash pickup by reducing my volume of trash to manageable levels. The few cans I use are flattened before disposal, and jars are reused when possible. Plastic containers, mixed paper and plastic packaging still contribute more to the volume than I like. We need to develop local sources for efficiently recycling plastic.

Feed garbage to worms. Because worms eat my garbage (thereby eliminating the materials that make the odor from mixed trash unbearable), the plastic bags in which I store trash are neither unpleasant, nor an attraction for varmints. I stockpile these in the garage, and combine a friend's trash with mine to make a dump run only three to four times a year. This costs less than $10 a year for both households involved.

The key to this system is the worms. Organic waste isn't mixed with the trash. The compactor-transfer station I use does not accept food waste from residents who haul their own trash, anyway. If I didn't have worms processing my garbage, I would have to pay extra for residential pickup by private haulers. (It also helps that I have a suitable vehicle for hauling trash, and space to stockpile it.)

The economic advantage of using worms to process kitchen waste can also be calculated in other ways.

Purchasing tagged bags. In some communities, residents pay for color-coded trash bags that are set out as necessary for weekly pickup. This method provides an incentive for lower volume, since the fewer bags put out, the fewer you have to pay for. One woman who installed a home vermicomposting system found that she only had to put out a trash bag about every third week, instead of weekly as she had been doing. It wasn't the volume that made the difference, but the fact that without the organic waste in the bag, she could stand the odor when opening it to deposit waste over a longer period of time.

One-can, two-can, three-can pickup. Differential rates for trash pickup are becoming more common among private haulers. If a three-garbage can family can install a home vermicomposting system, recycle all recyclables, and cut back to being a one-can family, they will save money. *And*, the landfill will last longer. *And*, it will be less likely to develop toxic leachates as organic acids from the garbage react with metals and other materials in the dump. *And*, the household will save money. *And*, the household pays less for fertilizer for its garden, *and* it spends less to purchase bait to go fishing, *and . . . and . . . and*. Why isn't everybody doing it?

Summary

Earthworms play an important role in recycling organic nutrients from dead tissues back to living organisms. They do this without fanfare; rarely does anyone see them perform their tasks.

If you decide to use worms to process your own organic kitchen waste, you will see them at work. You will see mounds of disagreeable material converted noiselessly, with almost no odor, to materials you can use directly on your houseplants and in your garden. You will enjoy healthier looking plants, better tasting vegetables, and will spend less on fertilizers and trash-disposal. Some of you will have fish on the table, attracted by the worms you have for bait. Hopefully, you'll also gain a better appreciation of the intricate balance and interdependencies in nature. You will be treading more gently upon the earth.

As your gardens are enriched, so is your life and mine. You will have joined the adventurers who say, "Worms eat my garbage," and isn't that a grand beginning on a task that needs to start somewhere? You, personally, can make it happen.

Afterword:
How many worms in an acre?

One year I counted and weighed all the earthworms I could handsort from the top seven inches of a square foot of my garden. I counted 62 worms of all sizes, and at least two species. If I had had an acre under cultivation and if this was, in fact, a representative sample, the total population would have exceeded 2.7 million worms per acre!

These 62 worms weighed two ounces. Extended to one acre, this would give a total weight of 5445 pounds, or over two and one-half tons of worms in the top seven inches of one acre (43,560 sq. ft.) of soil!

Earthworms of these soil dwelling types eat soil in their search for organic nutrients. This soil is mixed with the organic materials and bacteria in their intestines and is deposited as castings. The weight of these castings per worm per day could easily equal the weight of the worm. To take a conservative figure, let's estimate that the weight of castings deposited per day from one worm is one-eighth the weight of the worm. The total weight of castings produced per acre per day would be 680 pounds. Think of the value to the plants of those castings, and the activity of the worms in producing those castings.

To get an estimate of annual casting production, let's assume that the worms are active only 150 days of the year, giving you 102,093 pounds per year, or over 51 tons of castings per year. (If you have ever tossed a ton of manure onto, and off, a pickup, you can begin to appreciate the work worms perform for you in your garden.)

These calculations compare favorably with estimates from scientists around the world. In 1881, Darwin estimated that earthworms deposited from 7-1/2 to 18 tons of casts per acre in pasture. Stockli's work in 1928 comes closer to the figure of 51 tons per acre per year. He estimated that worms produced from 33 to 44 tons per acre in Switzerland.

Estimates from the rich Nile Valley are almost unbelievable. Beaugh estimated that earthworms deposit over 1000 tons of casts per acre per year. No other area in the world could be expected to exceed the casting production achieved by earthworms in the unusually favorable conditions of the Nile Valley.

The use of chemical fertilizers and pesticides has not only reduced soil earthworm and microbial populations, but also the amounts of organic matter present in the soil. These energy-intensive practices have led us to the point where applying greater quantities of fertilizers and pesticides at great expense does less good than the smaller quantities previously did. North temperate climatic regions, especially, could benefit from increased earthworm populations in soils. It is vital to help people understand the advantages of encouraging agricultural practices that increase native earthworm populations in our soils, and to discourage practices that kill them.

Glossary

acid Normal product of decomposition. Redworms do best in a slightly acid (pH just less than 7) environment. Below pH 5 can be toxic. Addition of pulverized egg shells and/or lime helps to neutralize acids in a worm bin. See pH.

aggregation Clustering, as of soil particles to form granules that aid in aeration and water penetration.

aeration Exposure of a medium to air to allow exchange of gases.

aerobic Pertaining to the presence of free oxygen. Organisms that utilize oxygen to carry out life functions.

albumin A protein in cocoons that serves as a food source for embryonic worms.

alkaline Containing bases (hydroxides, carbonates) which neutralize acids to form salts. See pH.

anaerobic Pertaining to the absence of free oxygen. Organisms that can grow without oxygen present.

anaerobiosis Life in an environment without oxygen or air.

anterior Toward the front.

ardox nails Nails with a spiral shape designed to increase holding power.

bedding Moisture retaining medium used to house worms.

bio-degradable Capable of being broken down into simpler components by living organisms.

biological control Management of pests within reasonable limits by encouraging natural predator/prey relationships and avoiding use of toxic chemicals.

biomass That part of a given habitat consisting of living matter, expressed as weight of organisms per unit area. Recommended biomass of worms for vermicomposting is about 1/4 to 1/2 lb per square foot surface area of bedding.

breeders Sexually mature worms as identified by a clitellum.

buffer A substance which renders a system less sensitive to fluctuations between acidity and alkalinity. Humus serves as a buffer in soil.

calcium carbonate Used to reduce acidity in worm bins and agricultural soils. See lime.

Canadian peat moss. See peat moss.

castings See worm castings.

CDX plywood CD plywood has knotholes and small splits present, as contrasted with a higher grade such as AB which has one side smooth and free from defects. Exterior (X) plywood is bonded with waterproof glue and suitable for use outside.

cellulose   An inert compound containing carbon, hydrogen and oxygen and a component of worm beddings. Wood, cotton, hemp and paper fibers are primarily cellulose.

chlorosis   Abnormal yellowing of plant tissues caused by nutrient deficiency or activities of a pathogen.

clitellum   A swollen region containing gland cells which secrete the cocoon material.

cocoon   Structure formed by the clitellum which houses embryonic worms until they hatch.

compactor-transfer station   A facility which accepts solid waste and compacts it prior to transfer to a landfill or other refuse disposal facility.

compost   Biological reduction of organic waste to humus. Used to refer to both the process and the end product. One composts leaves, manure, garden residues to obtain compost which enhances soil texture and fertility when used in gardens.

consumer   An organism that feeds on other plants or animals.

culture   To grow organisms under defined conditions. Also, the product of such activity, as a bacterial culture.

cyst   A sac, usually spherical, surrounding an animal in a dormant state.

decomposer   An organism that breaks down cells of dead plants and animals into simpler substances.

decomposition   The process of breaking down complex materials into simpler substances. End products of much biological decomposition are carbon dioxide and water.

diatomaceous earth   Finely pulverized shells of diatoms used for insect control.

earthworm   A segmented worm of the Phylum Annelida, most of whose 3000 species are terrestrial.

egg   A female sex cell capable of developing into an organism when fertilized by a sperm.

egg case   See cocoon.

Eisenia foetida   Scientific name for most common redworm used for vermicomposting.

enchytraeids   Small, white, segmented worms common in vermicomposting systems.

enzyme   Complex protein which provides a site for specific chemical reactions.

excrete   To separate and to discharge waste.

feces   Waste discharged from the intestine through the anus. Manure.

fertilize   To supply nutrients to plants, or, to impregnate an egg.

genus   A category of classification grouping organisms with a set of characteristics more generalized than species characteristics.

girdle   See clitellum.

gizzard Region in anterior portion of digestive tract whose muscular contractions help grind food.

hatchlings Worms as they emerge from a cocoon.

heavy metal Dense metal such as cadmium, lead, copper, and zinc which can be toxic in small concentrations. Build up of heavy metals in garden soil should be avoided.

hemoglobin Iron-containing compound in blood responsible for its oxygen-carrying capacity.

humus Complex, highly stable material formed during breakdown of organic matter.

hydrated lime Calcium hydroxide. Do not use in worm bins. See lime.

inoculate To provide an initial set of organisms for a new culture.

leach To run water through a medium, causing soluble materials to dissolve and drain off.

leaf mold Leaves in an advanced stage of decomposition.

lime A calcium compound which helps reduce acidity in worm bins. Use calcium carbonate, ground limestone, egg shells, or oyster shells. Avoid caustic, slaked, and hydrated lime.

limestone Rock containing calcium carbonate.

litter (leaf) Organic material on forest floor containing leaves, twigs, decaying plants, and associated organisms.

Lumbricus rubellus Scientific name for a redworm species found in compost piles and soils rich in organic matter.

Lumbricus terrestris Scientific name for large burrow-dwelling nightcrawler.

macroorganism Organism large enough to see by naked eye.

microorganism Organism requiring magnification for observation.

monoculture Cultivation of a single species.

nematodes Small (usually microscopic) roundworms with both free-living and parasitic forms. Not all nematodes are pests.

nightcrawler A common name for *Lumbricus terrestris,,* a large, burrow-inhabiting earthworm.

optimal Most favorable conditions, such as for growth or for reproduction.

organic Pertaining to or derived from living organisms.

overload To deposit more garbage in a worm bin than can be processed aerobically.

pasteurize To expose to heat long enough to destroy certain types of organisms.

pathogen Disease producing organism.

peat moss Sphagnum moss which is mined from bogs, dried, ground, and used as an organic mulch. Although acidic, its light, fluffy texture and excellent moisture retention characteristics make it a good component for worm bedding.

perlite   A lightweight volcanic glass used to increase aeration in potting mixtures.

pH   An expression for degree of acidity and alkalinity based upon the hydrogen ion concentration. The pH scale ranges from 0 to 14, pH of 7 being neutral, less than 7 acid, greater than 7, alkaline.

pharynx   Muscular region of the digestive tract immediately posterior to a worm's mouth.

pit-run   Worms of all sizes, as contrasted with selected breeders.

population density   Number of specific organisms per unit area, e.g. 1000 worms per square foot.

posterior   Toward the rear, back or tail.

potting soil   A medium for potting plants.

pot worms   See enchytraeids.

prostomium   Fleshy lobe protruding above the mouth.

protein   Complex molecule containing carbon, hydrogen, oxygen and nitrogen; a major constituent of meat. Worms are approximately 60% protein.

putrefaction   Anaerobic decomposition of organic matter, especially protein, characterized by disagreeable odors.

redworms   A common name for *Eisenia foetida* and also *Lumbricus rubellus*. *Eisenia foetida* is the primary worm used for vermicomposting.

regenerate   To replace lost parts.

run-of-pit   See pit-run.

salt   Salts are formed in worm bins as acids and bases combine, having been released from decomposition of complex compounds.

secrete   To release a substance that fulfills some function within the organism.

segment   One of numerous disc-shaped portions of an earthworm's body bounded anteriorally and posteriorally by membranes.

seminal fluid   Fluid which contains sperm that are transferred to an earthworm's mate during copulation.

setae   Bristles on each segment used in locomotion.

sexually mature   Possessing a clitellum.

shredded corrugated   By-product of carton manufacture processed to make worm bedding.

side dressing   Application of nutrients on soil surface away from stem of plants.

slaked lime   Calcium hydroxide. Do not use in worm bins.

species   Basic category of biological classification, characterized by individuals which can breed together.

sperm   Male sex cells.

sperm-storage sacs   Pouches which hold sperm received during mating.

subsoil   Mineral bearing soil located beneath humus-containing topsoil.

taxonomist   A scientist who specializes in classifying and naming organisms.

top dressing   Nutrient-containing materials placed on the soil surface around the base of plants.

toxic   Poisonous, life-threatening.

toxoplasmosis   Disease caued by the protozoan *Toxoplasmosis gondii*.

vermicompost   Mixture of partially decomposed organic waste, bedding, worm castings, cocoons, worms and associated organisms, or, to carry out composting with worms.

vermiculite   Lightweight potting material produced through expansion of mica by means of heat.

vermiculture   The raising of earthworms under controlled conditions.

white worms   See enchytraeids.

worm bin   Container designed to accommodate a vermicomposting system.

worm casting   Undigested material, soil, and bacteria deposited through the anus. Worm manure.

worm:garbage ratio   Relationship between weight of worms set up in a bin to process a given amount of garbage.

Metric conversions

To convert:
- pounds (lb) to kilograms (kg), multiply by 0.45
- inches (in., or ") to centimetres (cm), multiply by 2.54
- feet (ft, or ') to centimetres (cm), multiply by 30.5
- Farenheit degrees (F°) to Celsius (C°), subtract 32, then multiply by 5/9
- square feet (sq ft, or ft²) to square metres (m²), multiply by 0.09

page	line	English unit	Metric conversion
viii	1	1 lb.	0.5 kg
viii	1	65 lb.	29.3 kg
4	1	50°F.	10°C
4	2	55-77°F.	13-25°C
4	3	84°F.	29°C
10	7	8 to 12 in.	20 to 30cm
11	18	5, 10, 15 lb.	2.3, 4.5, 6.8 kg
12	3	2 ft x 2 ft x 8 in.	60x60x20 cm
12	4	sq ft/pound/week.	0.2 m²/kg/week
12	7	1.75 to 12 lb.	0.8 to 5.4 kg
12	9	5.2 lb.	2.3 kg
12	10	1 ft x 2 ft x 3 ft.	30x60x90 cm
12	11	6 sq ft.	0.54 m²
12	30	1' x 2' x 3'.	30x60x90 cm
12	30	2' x 2' x 8".	60x60x20 cm
12	32	6 lb.	2.7 kg
12	36	⅝" x 35⅝" x 12".	1.6 x 88.4 x 30 cm
12	37	⅝" x 23⅜" x 12".	1.6 x 58.4 x 30 cm
12	38	⅝" x 24" x 36".	1.6 x 60 x 90 cm
12	39	2", ½".	5 cm, 1 cm
13	4	½".	1 cm
14	12	2' x 2' x 8".	60 x 60 x 20 cm
14	14	⅝" x 23⅜" x 8".	1.6 x 58.4 x 20 cm
14	15	⅝" x 24" x 24".	1.6 x 60 x 60 cm
14	16	2", ½".	5 cm, 1 cm
17	27	1 to 3 inches.	2.5 to 7.6 cm
19	34	½".	1 cm
23	10	50°F.	10°C
26	34	½ to 1 inch.	1 to 2.5 cm
26	36	1 lb.	0.5 kg
28	footnte.	77°F.	25°C

page	line	English unit	Metric conversion
30	24	7 lb.	3.2 kg
30	25	1 lb.	0.5 kg
30	26	½ lb.	0.2 kg
30	28	1 lb to ½ lb.	0.5 to 0.2 kg
31	2	4 sq ft.	0.4 m²
31	3	1 sq ft per lb.	0.2 kg/m²
31	15	1000/lb.	2200/kg
31	18	600-700/lb.	1300-1600/kg
31	23	150,000-200,000/lb. . . .	300,000-400,00/kg
31	25	2000/lb.	4000/kg
32	38	10°F, 90°F.	−12°C, +32°C
34	29	12 lb to 4 lb.	5.4 to 1.8 kg
35	4	4 to 6 lb.	1.8 to 2.7 kg
35	5	9-14 lb.	4.1-6.3 kg
37	11	80-90°F.	27-32°C
43	2	1 in.	2.5 cm
61	7	¼ to 1 in.	0.6 to 2.5 cm
61	26	⅟₁₆ in.	0.2 cm
64	5	¼".	0.6 cm
81	6	2.7 million/acre.	6.75 million worms/hectare
81	7	2 oz.	56.6 g
81	8	5445 pounds.	2450 kg
81	9	2½ tons/acre.	5.6 tonnes per hectare
81	19	680 pounds.	765 kg per hectare
81	23	102,093 pounds.	45,942 kg
81	23	51 tons per acre.	115 tonnes per hectare
82	1	7½ to 18 tons/acre.	16.8 to 40.5 tonnes per hectare
82	3	51 tons per acre.	115 tonnes per hectare
83	4	33 to 44 tons per acre . . .	75 to 100 tonnes per hectare
82	6	1000 tons per acre	2600 tonnes per hectare

Record sheet

Date set up _____ = Day 0

Description of set-up:

Initial weight of worms _____

☐ Breeders or ☐ mixed sizes

Type of bedding _____

Size of bin_____

Number in household _____

Garbage burying locations:

	2'				3'			
	1	6	7		1	6	7	12
2'	2	5	8	2'	2	5	8	11
	3	4	9		3	4	9	10

Date	Day	# oz.	Total # oz. to date	Temp.	Water # of pints	Burying location #	Com- ments

Date harvested _____ No. of Days _____ Worm wgt. _____

Total wgt. garbage buried _____ oz. = _____ lb.

Wgt. uneaten garbage _____

Ave. oz. buried per day _____

Ave temp. _____ Temp. range _____

Suggested References

Earthworms

Lauber, Patricia. *Earthworms, Underground Farmers.* Champaign, IL: Garrard Publishing Co. 1976. 64p
This colorful, informative, technically accurate book presents earthworm structure, ecology and physiology in language a third-grader could understand. It's generous use of color photographs of worms and their natural enemies surpasses that found in any other publication known to this author.

Hopp, Henry. *What Every Gardener Should Know About Earthworms.* Charlotte, VT: Garden Way Publishing Co. 1973. 39p
Although much of the information in this meaty little booklet was adapted from a 1954 publication entitled *Let An Earthworm Be Your Garbage Man,* the presentation on effects of earthworms on soil moisture, aeration, and soil fertility is still pertinent.

Sroda, George. *No Angle Left Unturned: Facts About Nightcrawlers.* Amherst Junction, WI: George Sroda. 1975. 111p
Written by a man who has studied nightcrawlers for years, this practical manual tells how to harvest, hold, feed, water, and condition them for fishing.

Minnich, Jerry. *The Earthworm Book.* Emmaus, PA: Rodale Press. 1977. 372p
Written for the lay-person, this readable book traces the earthworm through history, its environment, and its use as a soil builder, composter, and income-producer. The final chapter reprints case histories of people who have raised earthworms for profit or soil-building purposes.

Edwards, C.A., and J.R. Lofty. *Biology of Earthworms.* 2nd ed. London: Chapman and Hall. Available in U.S. from Methuen, Inc., New York, NY. 1977. 333p
More technical than the preceeding books, this newly revised text covers morphology, taxonomy, biology, physiology, ecology, the role of earthworms in organic matter cycles, and other subjects related to earthworms. Written by scientists currently conducting earthworm research, this book is a must for serious students of earthworms.

Appelhof, Mary (compiler). *Workshop on the Role of Earthworms in the Stabilization of Organic Residues. Vol. I: Proceedings.* Kalamazoo, MI: Beech Leaf Press of the Kalamazoo Nature Center. 1981. 340p

World's leading investigators of earthworms review pertinent information, present current data, and project future research needed to develop the potential for organic waste conversion by means of earthworms in twenty-eight papers presented at a major research-needs workshop supported by the National Science Foundation. Provides the most up-to-date information available from academic, commercial, and governmental viewpoints. Accompanied by a companion, *Vol. II: Bibliography*, compiled and edited by Diane D. Worden in 1981, 492p, which provides access to journal articles, books, patents, dissertations, and federally supported research. 3036 citations are fully indexed.

Earthworm Farming

Shields, Earl B. *Raising Earthworms for Profit.* Eagle River, WI: Shields Publications, P.O. Box 669, Eagle River, WI. Original copyright, 1959, 17th edition, 1978. 128p.

This manual has been the standard training device for hundreds if not thousands, of worm growers. It discusses markets, propagation boxes, indoor and outdoor pits, feeds, packing and shipping, and advertising. The basic text was written in the 1950's. Although minor revisions since then help to keep the content up-to-date, the 1978 edition says, "A fast worker, on a 'piece-work' basis, may earn as much as $1.25 per hour." Hardly minimum wage for these times.

Barrett, Thomas J. *Harnessing the Earthworm.* Wedgwood Press. Available from Shields Publications. 1947, 1959. 166p

Frustrating because it lacks a bibliography, this important document synthesizes much of the early literature on the effects of earthworms on soil fertility. Discusses humus, topsoil, subsoil, earthworm tillage, and chemical composition of earthworm castings. Excellent information for earthworm culture is provided.

Gaddie, Ronald E., Sr., and Donald E. Douglas. *Earthworms for Ecology and Profit. Vol. I: Scientific Earthworm Farming.* 1975. 180p. *Vol. II: Earthworms and the Ecology.* 1977. 263p Ontario, CA: Bookworm Publishing Co.

Full of information, and with some misinformation, these books are best used by someone able to wade through the poor editing and discriminate between what is really true, and what is only partially true. They do present more up-to-date information on commercial earthworm growing than previous works mentioned.

Composting

Dindal, Daniel L. *Ecology of Compost: A Public Involvement Project.* Syracuse, NY: State University of New York College of Environmental Science and Forestry. 1972. 12p (25¢)

A soil ecologist presents a primer on outdoor composting; discussing energy sources, decomposition rates, the carbon-nitrogen ratio, moisture, aeration, and heat production. He described the relationships between the organisms found in "Food Web of the Compost Pile" which is used in this book. A bargain for the price!

Minnich, Jerry, and Marjorie Hunt. *The Rodale Guide to Composting.* Emmaus, Pennsylvania: Rodale Press, Inc. 1979. 405p

A comprehensive, readable book which gives history, benefits, techniques, materials, and machines related to composting. Individuals who want to know everything about compost should read this book.

Soil Animals

Schaller, Friedrich. *Soil Animals.* Ann Arbor: University of Michigan Press. 1968. 144p

If you were enticed by the chapter on "other critters," but are not yet ready for a zoology text, look at this well-illustrated little book describing collection methods, characteristics, importance, habits, and behavior of animals that live in the soil.

Recycling and the environment

Goldstein, Jerome. *Recycling.* New York: Schocken Books. 1979. 238p

Taking the position that recycling begins at home, Goldstein describes what individuals have done to recycle their waste, then extends his discussion to community, municipal, and industrial recycling programs. He discusses "bottle bills," public policies, large-scale composting, and looks for a future without dumps.

Geller, E. Scott, Richard A. Winett, and Peter B. Everett. *Preserving the Environment: New Strategies for Behavior Change.* Elmsford, NY: Pergamon Press. 1982. 338p

Recognizing that most individuals think physical technology will alleviate our problems, three psychologists describe over 150 studies carried out over the past decade which demonstrate many cases where changes in behavior will do more than technology to enhance the quality of our lives. In academic terms they say of vermicomposting: "The response-maintenance reinforcers have included the convenient availability of plant fertilizer and fishing worms, and special social attention (everyone wants to see the 'innovative' worm bin!)."

Sources

Appelhof, Mary. "Basement worm bins produce potting soil and reduce garbage." Kalamazoo, MI: Flowerfield Enterprises, 1973. 2p

Appelhof, Mary. "Composting your garbage with worms," Kalamazoo, MI: Kalamazoo Nature Center, 1979. Revised 1981. 4p

Appelhof, Mary. "Household scale vermicomposting," in *Workshop on the Role of Earthworms in the Stabilization of Organic Residues. Vol. I: Proceedings*, compiled by Mary Appelhof. Kalamazoo, Michigan: Kalamazoo Nature Center Beech Leaf Press, 1981. p232-240

Appelhof, Mary. "Vermicomposting on a household scale," in *Soil Biology as Related to Land Use Practices, Proceedings of the International Colloquium on Soil Zoology*, edited by Daniel Dindal. U.S. EPA, 1980. p157-160

Appelhof, Mary. "Winter composting with worms," Final report to National Center for Appropriate Technology. Kalamazoo, MI: Kalamazoo Nature Center, 1979. 13p

Appelhof, Mary. "Worms—a safe, effective garbage disposal," *Organic Gardening and Farming*, 21:8 (1974) p65-69

Appelhof, Mary. "Worms vs. high technology," *Creative Woman*, 4:1 (1980) p23-28

Appelhof, Mary, Michael Tenenbaum, and Randy Mock. "Energy considerations: Resource recycling and energy recovery," Presentation before the Resource Recovery Advisory Committee, South Central Michigan Planning Council, July 1980.

Appelhof, Mary, Michael Tenenbaum, Randy Mock, Cheryl Poché, and Scott Geller. *Biodegradable Solid Waste Conversion into Earthworm Castings*, Final report to National Science Foundation ISP-8009755. Kalamazoo, MI: Flowerfield Enterprises, 1981 78p

Barrett, Thomas J. *Harnessing the Earthworm*. Boston, MA: Wedgwood Press, 1959, original edition 1947. 166p

Beauge, A. "Les vers de terre et la fertilité du sol," *J. Agric. prat Paris* 23 (1912) 506-507, Use of this material is from C.A. Edwards, *Biology of Earthworms*, 2nd ed, p144

Darwin, Charles, *The Formation of Vegetable Mould, through the Action of Worms, with Observations on their Habits.* New York: D. Appleton and Company, 1898 from 1881 edition, 326p

Dindal, Daniel L. "Ecology of Compost: A Public Involvement Project." Syracuse, New York: NY State Council of Environmental Advisors and the State University of New York College of Environmental Science and Forestry, 1972. 12p

Edwards, C.A. and J. R. Lofty. *Biology of Earthworms*, 2nd edition. London, United Kingdom: Chapman and Hall, 1977. 333p

Geller, E. Scott, Richard A. Winett, and Peter B. Everett. *Preserving the Environment*. Elmsford, New York: Pergamon, 1982. 338p

Goldstein, Jerome. *Recycling*. New York: Schocken Books, 1979. 238p

Handreck, Kevin Arthur. "Earthworms for Gardeners and Fishermen." Adelaide, Australia: CSIRO Division of Soils, 1978. 15p

Hartenstein, R., E.F. Neuhauser and J. Collier. "Accumulation of heavy metals in the earthworm *Eisenia foetida*," *Journal of Environmental Quality*, 9 (1980) 23-26

Hartenstein, R., E.F. Neuhauser, and D.L. Kaplan. "Reproductive potential of the earthworm *Eisenia foetida*," *Oecologia* 43 (1979) 329-340

Home, Farm and Garden Research Associates. *Let an Earthworm Be Your Garbage Man*, Eagle River, WI: Shields, 1954. 46p

Kaplan, D.L., R. Hartenstein, E.F. Neuhauser. "Coprophagic relations among the earthworms *Eisenia foetida, Eudrilus eugeniae* and *Amynthas* spp," *Pedobiologia*, 20 (1980), p74-84

Kaplan, D.L., E.F. Neuhauser, R. Hartenstein and M.R. Malecki. "Physicochemical requirements in the environment of the earthworm *Eisenia foetida*," *Soil Biology and Biochemistry*, 12 (1980), 347-352

Martin, J.P., J.H. Black, and R.M. Hawthorne. "Earthworm Biology and Production." University of California Cooperative Extension leaflet # 2828, 1976. 10p

McCormack, Jeffrey H. "A review of whitefly traps," *The IPM Practitioner*, 3:10 (1981) p3

Minnich, Jerry and Marjorie Hunt. *The Rodale Guide to Composting*. Emmaus, PA: Rodale Press, 1979. 405p

Mitchell, Myron J., Robert M. Mulligan, Roy Hartenstein, and Edward F. Neuhauser. "Conversion of sludges into 'topsoils' by earthworms," *Compost Science*, Jul/Aug (1977) 28-32

Morgan, Charlie. *Earthworm Feeds and Feeding*, 6th edition. Eagle River, WI: Shields, 1972. 90p

Munday, Vivian, and J. Benton Jones, Jr. "Worm castings: How good are they as a potting medium?" Southern Florist and Nurseryman 94:2 (1981), 21-23

Neuhauser, Edward F., Roy Hartenstein and David L. Kaplan. "Second progress report on potential use of earthworms in sludge management," in *Proceedings of Eighth National Conference on Sludge Composting*. Silver Springs, MD: Information Transfer, Inc., 1979. p238-241

Neuhauser, E.F., D.L. Kaplan, M.R. Malecki and R. Hartenstein. "Materials supporting weight gain by the earthworm *Eisenia foetida* in waste conversion systems," *Agricultural Wastes* 2 (1980), 43-60

Myers, Ruth. *A Worming We Did Go!* Elgin, IL: Shields, 1968. 71p

Reynolds, John W. *The Earthworms (Lumbricidae and Sparganophilidae) of Ontario*. Toronto, Canada: Royal Ontario Museum, 1977. 141p

Satchell, John E. "Earthworm evolution: Pangaea to production prototype," in *Workshop on the Role of Earthworms in the Stabilization of Organic Residues. Vol. I: Proceedings*, compiled by Mary Appelhof. Kalamazoo, MI: Kalamazoo Nature Center Beech Leaf Press, 1981. p3-35

Satchell, John E. "Lumbricidae," in *Soil Biology*, edited by A. Burges and F. Raw. London and New York: Academic Press, 1967. p259-322

Schaller, Friedrich. *Soil Animals*. Ann Arbor, MI: University of Michigan Press, 1968. 144p

Stockli, A. "Studien über den Einfluss der Regenwurner auf die Beschaffenheit des Bodens," *Landw. Jb. Schweiz.* 42 (1928):1, Use of this material is from C.A. Edwards, *Biology of Earthworms*, 2nd ed. p144

Vick, Nicholas A. "Toxoplasmosis," in Grinker's Neurology, 7th edition.

Worden, Diane D., editor, *Workshop on the Role of Earthworms in the Stabilization of Organic Residues, Vol. II: Bibliography*. Kalamazoo, MI: Kalamazoo Nature Center Beech Leaf Press, 1981. 492p